THE MATRIX ALGEBRA CALCULATOR:

LINEAR ALGEBRA PROBLEMS
FOR COMPUTER SOLUTION

CHARLES H. JEPSEN

EUGENE A. HERMAN

Grinnell College

BROOKS/COLE PUBLISHING COMPANY

Pacific Grove, California

Brooks/Cole Publishing Company
A Division of Wadsworth, Inc.

Printed in the United States of America

10 9 8 7 6 5 4 3 2 1

ISBN 0 534 09300 0

Sponsoring Editor: *Jeremy Hayhurst*
Production Editor: *Penelope Sky*
Cover Design: *Roy Neuhaus*

Preface

This is a book of linear algebra problems designed for solution using MAX – the MAtriX Algebra Calculator software package. Part of the book is therefore devoted to an explanation of how to use the software. Most of it, however, is organized into three chapters of fifteen sections that cover the topics of a typical linear algebra course in a standard order.

The treatment within these sections is not standard. Because the book is meant to accompany, not replace, a standard linear algebra textbook, it consists largely of examples and problems that involve the use of a computer. Definitions and facts are introduced only when they are needed, and proofs are given only when they are short and instructive.

To us, using the computer in linear algebra seems natural, even long overdue. The computer, in a way, is responsible for the current undergraduate course. Forty years ago, linear algebra was a graduate mathematics course. Then the computer came along and made matrix calculations practicable, which encouraged scientists from all disciplines to create and exploit mathematical models based on linear algebra and requiring ever more extensive matrix computations. The demand for the current course is in large part a response to the increased need which these applications created for linear algebra in the sciences. Thus, applications to the sciences, including applications requiring a computer, ought to be a part of the course.

A good way to comprehend the benefits of using the computer in linear algebra is to work through our problems; they are the heart of the book. Computational problems, which occupy the bulk of the student's time in a standard course, are usually tedious and easily ruined by arithmetic errors. Here they are quick and reliable, freeing the student for more instructive activities. Experimentation and exploration problems are not even feasible in a standard course, since examples take too long to work through. Here we encourage the student to look for patterns, compare alternative methods, and pursue variations of a basic problem by asking what-if questions. Applications problems in a standard course are usually oversimplified to make the matrices small in size with small integer entries, or they are kept abstract. Here they are concrete and moderately realistic.

The matrices needed for these problems – all 160 of them – are available from within MAX, which relieves the user from having to enter data. They are a varied collections of matrices, some with mathematically significant characteristics and some containing data from applications problems, some so small that they can be manipulated by hand and some so large that hand computation would be unthinkable. Of course, other matrices can be entered from the keyboard or a file.

Because of our emphasis on the role of the computer, we discuss a few topics not commonly covered in linear algebra courses. These include LU factorization in Chapter 1 and QR factorization in Chapter 2. The most substantial addition, however, is the large and varied collection of applications in every chapter, made accessible by the power of the computer. These span such disparate topics as temperature distributions on a grid, a Leontief input/output model, equations for chemical reactions, graphs and directed graphs, bivariate and multi-variate least-squares data from several disciplines, Markov chains, a Leslie population model, multi-spring and multi-tank problems, and biological transport problems.

The book and the software have both been several years in the making, and earlier versions have been used by numerous colleagues. As a result, we have many to thank. For financial support, we thank the Alfred P. Sloan Foundation and its program in The New Liberal Arts, Grinnell College, and Brooks/Cole Publishing Company. For her expert use of TeX and her dedication to professional quality, our secretary Angela Johnson. For their uncommon proficiency in programming and their insights into program design, former Grinnell students John Norman (version 1.1), Gregory Banik (version 2.0), Lorenz Huelsbergen (version 2.1), and Albert Goodman (version 3.1). For inviting us to try out early versions of our materials on colleagues, the Minicourse Committee of the Mathematical Association of America. For their helpful comments on these materials, many of the one hundred or more colleagues who used them, especially James Hurley of the University of Connecticut, David Kraines of Duke University, and Rudolf Lidl of the University of Tasmania. And, finally, we thank Christopher Jepsen for his help in preparing some of the data.

We also acknowledge the books that influenced us and the major software packages we used. The numerical algorithms in MAX came from the LINPACK and EISPACK projects at Argonne National Laboratory. Our exposition and problems were most influenced by the books in our References list by Arganbright, Helzer, Rorres and Anton, Strang, and Tucker. We developed our software on a PDP 11/70 and VAX 8600 using Oregon

Software's Pascal compiler and then ported it to Microsoft's Pascal compiler on an MS-DOS system. The book was produced with TeX.

Note to the Instructor

It is not a trivial matter to make effective use of supplementary materials and even less so when these materials include a software package. We have found that for students to benefit substantially from our materials, we must incorporate them into the course just as we do a textbook. That is, we discuss them in class, assign many of the problems, discuss some of the problems, and give relevant exam questions.

In fact, the computer has had such a strong effect on how we think about linear algebra that we have also altered the emphasis in the course. For example, we now spend less time on methods of hand computation and on determinants, but more time on applications and on eigenvalues and eigenvectors.

Fortunately, little time needs to be spent learning to use the software. Before the first computer assignment, we demonstrate MAX for the students, and later in the semester we answer occasional questions about it.

Assignments and exam questions require more effort. We like to sprinkle the computer assignments over much of the semester, so students realize that these are an essential and continuing part of the course. To keep the routine computer problems from being overly routine, we often ask the students to show on their papers a derivation of the group storage matrix, and we often assign problems where students have to figure out how to modify an existing matrix. Our computer-related exam questions usually are one of the following two types. Output from MAX is displayed and the student must answer questions or do further computations based on it. Or, the student is asked to show how to use the program to get a specified result and to explain what the output would show.

More suggestions can be found in the report from a MAA panel discussion [Herman et al (to appear in 1987)].

CONTENTS

An Overview of MAX

MAX, the MAtriX Algebra Calculator, is designed to be both powerful and easy to use. It can solve a system of equations, find a basis of a vector space, invert a matrix, factor a matrix, find a complete set of eigenvalues and eigenvectors of a matrix, compute an arbitrary algebraic expression involving matrices, and more. Other of its capabilities make it easier for you to construct and organize the matrices that you want to perform such computations on.

To find out how to get into the program, see the instructions accompanying the diskette; more extensive instructions appear in Appendix IV. You will know that you are in MAX when you see on the screen

```
MAX -- the MAtriX Algebra Calculator (version 3.1)


               Copyright 1987
               Wadsworth, Inc.
```

Command:

You respond to this and subsequent prompts by giving commands, as we explain below. To get out of MAX, you give the **exit** command.

This overview is intended to help you understand the MAX program as a whole and learn the basic responses to its **Command:** prompt. When you have read this section and tried out the examples, you will be prepared to begin reading Chapter 1 and doing the problems it presents. Another way to get started is to run through one of the demonstrations that accompanies the program. For example, you can give the command

Demonstrate "Demo1"

when you have gotten into the program; then you press the RETURN key as each command appears on the screen.

Before you can command MAX to perform a computation on a matrix, you must get that matrix into your <u>workspace</u>. This workspace is a temporary storage area that can hold many matrices of various sizes, that only you have access to, and that loses all its contents when you exit the program (see the figure). Normally, you will get your matrices into the workspace from <u>group storage</u>, although you can also enter them from the keyboard. Group storage, in contrast to the workspace, is a permanent storage area that contains even more matrices, that everyone has access to, and that remains intact when you exit the program. The matrices in group storage are, in fact, exactly the matrices that you need to solve the problems in this book. They are arranged in a list, numbered 1-200 (with some gaps), and are referred to in the problems within square brackets. Thus, [25] refers to matrix 25 in group storage.

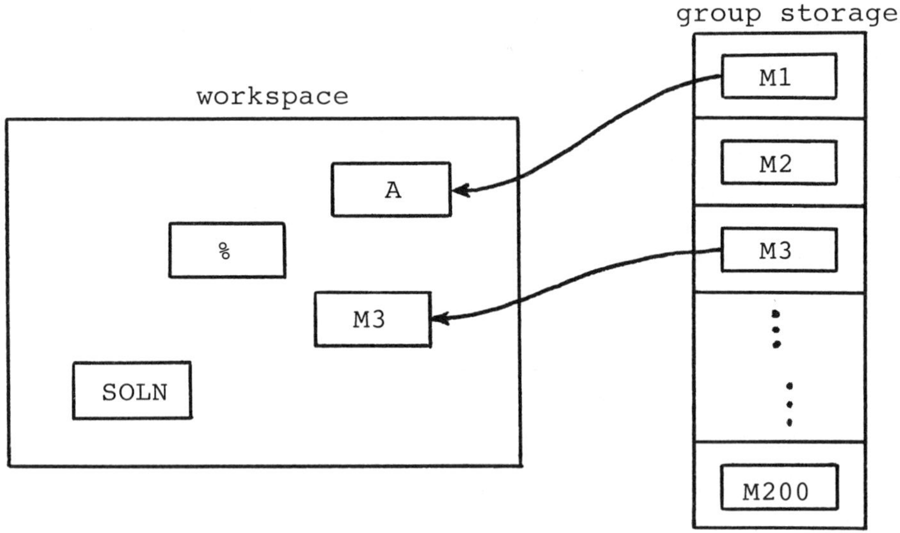

As the figure suggests, every matrix in your workspace has to have a <u>name</u>, although MAX can assign default names for you. Names that you assign must have between 1 and 10 alphanumeric characters and the first character must be a letter. However, there are two names that MAX will not let you assign to a workspace matrix because they are used only in special contexts. The name I may be used only in the **compute** command, where it always denotes an identity matrix of the appropriate size. The name ALL is used to refer to all the matrices in your workspace, as in **delete all**, which is a command to delete all the matrices in your workspace. The default name that MAX usually assigns when you do not name a matrix is %. Then the next matrix named % automatically overwrites the preceding one of that name. Thus, % is a good name for a matrix that you need to see only once and would rather not have cluttering up your workspace. A few commands result in two new workspace matrices, not just one; the default name of the first is % and the second is %%. On the other hand, the **get** command, which you use to get a matrix from group storage to the workspace, assigns the less forgettable default name M1 to the first matrix in group storage, the name M2 to the second matrix in group storage, M3 to the third, and so on.

There are two methods by which you can give commands to MAX. The <u>prompting</u> method is easier for beginners. First you enter a command name. These are mostly self-explanatory names such as **get, solve, multiply, display, augment,** and **exit.** When you press the RETURN key, MAX displays the form of the command and then prompts you for each of the command's <u>parameters</u>, which are the capitalized words following the command name. When you have responded successfully to all of these, MAX executes the command and displays the result.

If you forget the name of the command you want, you can see a list of all their names by giving the **help** command. Also, you can see a description of what a command does by typing **help** followed by the command name; for example, the command **help solve** directs MAX to show you a description of the **solve** command.

In the example below, we enter the command name **get** and then respond to the prompts for the parameters Integer1 and Matrix1 by entering **2** for the first parameter and accepting the default M2 for the second (by simply pressing RETURN). The result is that MAX gets a copy of the second matrix in group storage and stores it in our workspace under the name M2.

```
Command:  get
Get matrix Integer1 and store it in Matrix1
Integer1:  2
Matrix1: <M2>
Get matrix 2 and store it in M2

        2.        2.        1.
        3.        0.        4.
        1.       -1.        2.
```

```
Command:  transpose
Transpose Matrix1 and store the transpose in Matrix2
Matrix1:  m2
Matrix2: <%>
Transpose M2 and store the transpose in %
        2.        3.        1.
        2.        0.       -1.
        1.        4.        2.
```

Our second command causes MAX to find the transpose of M2 and to store this transpose in our workspace under the name %. If we had wanted to retain this transpose, we probably would have assigned it an alphanumeric name rather than accepting the default name %. You will recognize when a prompt has a default response; it is always displayed within angle brackets. By the way, MAX examines only the first three letters of a command name. Thus, in place of **transpose**, we could have typed **tra** or any other word that begins with those three letters.

MAX automatically displays the result of any command, if the result is a new or changed matrix in the workspace. Thus, after the above two commands, MAX displayed the result matrices M2 and the transpose of M2, respectively. Sometimes, however, we will want to speed up our session with MAX by suppressing some or all of these displays. To suppress the display of a single command's result, we type the special character # at the end of the command; to suppress all subsequent automatic displays, we give the **noshow** command.

The faster method by which you can give commands to MAX is the <u>full command</u> method. You can learn this way of communicating with MAX by imitating the echoes that it displays just before the result. For example, we could have obtained the same results as in the above example by giving the full commands

Get matrix 2 and store it in M2

Transpose M2 and store the tranpose in %

Full commands, however, are rarely written out in such detail. Except for the command names and the parameters, the other words are not needed. These filler words are ignored by MAX and are there to help you understand what the command does. Thus, we would normally give the above full commands in the abbreviated form

get 2

tra m2

Note that we did not need to include names for the two result matrices, since we were willing to accept the default names M2 and %, respectively.

If you do want to assign a name to the result, just add it to the end of the full command. For example, **get 2 A** assigns the name A to the result matrix. But if a command produces two result matrices, you must enclose their names in square brackets and separate them by one or more spaces, as in the next example.

```
Command:  get 2 a
Get matrix 2 and store it in A
     2.          2.          1.
     3.          0.          4.
     1.         -1.          2.
Command:  lufactor a [al au]
LUfactor A and store the L and U factors in [AL AU]
The product of the following L and U factors is A with row swaps:
  1 <--> 2
     1.          0.          0.
     0.67        1.          0.
     0.33       -0.5         1.

     3.          0.          4.
     0.          2.         -1.67
     0.          0.         -0.17
```

When MAX detects an error in a command, it sounds a bell, gives an explanation, and tries to bracket the incorrect part of your command by carets ∧. Often when you have made an error, you will decide not to complete the command. To abort the command, enter **cancel** in response to a prompt for one of its parameters. Or, if the prompt has no default response, simply press the RETURN key. (It is therefore unwise to begin a matrix name with the letters CAN; in this context the matrix name is interpreted as a **cancel**.)

An especially useful and general feature of MAX is its ability to recognize <u>submatrix expressions</u>. For example,

M2(1-3,1-2)

denotes the submatrix of M2 consisting of its 1st through 3rd rows and 1st through 2nd columns. The general form of a submatrix expression is

$$(\text{First Row-Last Row, First Column-Last Column})$$

There are also abbreviations for many common submatrix expressions. For example,

$$\text{M2(,1-2) means M2(1-n,1-2)} \qquad \text{(if M2 has } n \text{ rows)}$$

$$\text{M2(2-3,2) means M2(2-3,2-2)}.$$

Also, the comma may be omitted if the matrix has only one row or only one column. Almost any command that acts on a workspace matrix can act on a submatrix, and any command that produces a result matrix can insert that result into a submatrix of an existing workspace matrix. In the following example, we use submatrix expressions to begin a step-by-step reduction of a matrix.

```
Command:  get 5 a
Get matrix 5 and store it in A
    3.          -1.          2.
    2.           1.          1.
    1.           3.          0.
Command:  compute

Compute Expression1 and store in Matrix1
Expression1:  -2/3*a(1,) + a(2,)
Matrix1: <%> a(2,)
Compute (((-2/3)*A(1-1,1-3)) + A(2-2,1-3)) and store in A(2-2,1-3)
    3.          -1.          2.
    0.           1.67       -0.33
    1.           3.          0.
```

You might have noticed some <u>round-off error</u> in the above examples. For instance, the (2,3) entry of the last result should be -1/3 rather than -0.33. Such errors are a part of MAX and can, on rare occasions, obscure a result. There is value in occasionally encountering a result that has been obscured by round-off error. You learn that even the best of the current algorithms for numerical matrix computations are imperfect, and you discover that for some matrices an associated numerical quantity may be close to a borderline between two significantly different values. However, you will rarely have serious questions about the numbers produced by MAX, unless your instructor assigns you problems exhibiting such borderline behavior. Few of the problems in this book will cause such difficulties. Furthermore, the algorithms used in MAX are highly reliable and accurate, and the computations are carried out to about seven places. Although only two decimal places are normally displayed on the screen, all seven are stored. You can see more of them by using the **decimal** command.

One command in which you will almost always have to make a judgment about round-off error is the **rank** command. MAX is designed to save you from tedious computations, not to save you from making judgments, and determining the rank of a matrix requires a judgment. The **rank** command produces an upper-triangular matrix with the same rank as the given matrix, and you must judge how many nonzero rows it has. In the following example, we judge that matrix A has rank 2, since the (3,3) entry of the upper-triangular matrix is almost certainly zero. (The number 2.38E-07 means $(2.38)10^{-7}$.)

```
Command:  get 4 a
Get matrix 4 and store it in A
    1.          2.          3.
    3.          2.          1.
    0.          2.          4.
Command:  rank a
Rank -- record the rank of A
The following upper triangular matrix has the same rank as A:
    -5.1        -1.18       -3.14
                -2.94       -1.47
                             2.38E-07
Determinant =  0.00E0      r-condition =        0.
Enter the rank of A (it is at most 3):  2
```

MAX then records the fact that matrix A has rank 2 and uses it in several other commands: **solve, qrfactor, orthonormalize, nbasis,** and **cbasis.**

All the numbers in the examples were generated on a VAX. On a PC, the numbers will be slightly different, especially the ones that are nearly zero.

For more detailed information on all the commands, see Appendix I.

Chapter 1

Matrices and Systems of Equations

Section 1.1 ALGEBRA OF MATRICES

We begin with basic matrix operations: addition, subtraction, scalar multiplication, matrix multiplication, transposition. All can be performed using MAX commands as shown in the following examples. (We remind you that the Overview describes how to think about the MAX package as a whole and how to get started interacting with the package.)

Example 1: Let A be matrix 1 in group storage, let B be matrix 2 in group storage, and compute the matrices $A + B$, $3B$, $A - 3B$, AB.

```
Command: get

Get matrix Integer1 and store it in Matrix1
Integer1: 1
Matrix1: <M1> a
Get matrix 1 and store it in A

     -5.          1.          5.
     -7.          4.          4.
     -1.          1.          1.

Command: get

Get matrix Integer1 and store it in Matrix1
Integer1: 2
Matrix1:  <M2> b
Get matrix 2 and store it in B

      2.          2.          1.
      3.          0.          4.
      1.         -1.          2.
```

Command: **add**

Add Matrix1 to Matrix2 and store the sum in Matrix3
Matrix1: **a**
Matrix2: **b**
Matrix3: <%> s
Add A to B and store the sum in S

-3.	3.	6.
-4.	4.	8.
0.	0.	3.

The full command version of this addition command is:

Command: **add a b**
Add A to B and store the sum in %

-3.	3.	6.
-4.	4.	8.
0.	0.	3.

Since we did not supply a name for the sum, MAX assigned the default name %.

Command: **multiply**

Multiply Matrix1 (or Scalar1) by Matrix2 (or Scalar2) and store
  in Matrix3
Matrix1 (or Scalar1): **3**
Matrix2 (or Scalar2): **b**
Matrix3: <%> c
Multiply 3. by B and store in C

6.	6.	3.
9.	0.	12.
3.	-3.	6.

Command: **subtract**

Subtract from Matrix1 Matrix2 and store the difference in Matrix3
Matrix1: **a**
Matrix2: **c**
Matrix3: <%> **d**
Subtract from A C and store the difference in D

-11.	-5.	2.
-16.	4.	-8.
-4.	4.	-5.

Command: **multiply**

Multiply Matrix1 (or Scalar1) by Matrix2 (or Scalar2) and store
    in Matrix3
Matrix1 (or Scalar1): a
Matrix2 (or Scalar2): b
Matrix3: <%> p
Multiply A by B and store in P

```
    -2.        -15.          9.
     2.        -18.         17.
     2.         -3.          5.
```

Again this can be performed with a single command:

Command: **multiply a b**
Multiply A by B and store in %

```
    -2.        -15.          9.
     2.        -18.         17.
     2.         -3.          5.
```

Once more MAX provided the default name % for an unnamed matrix. The
previous value of % has now been replaced by the new value.

Example 2: Let A be matrix 3 in group storage and compute $A'A$.

Note: Here and throughout this book, we use the symbol A' to denote the
transpose of A.

Command: **get 3 a**
Get matrix 3 and store it in A

```
     5.          0.          3.
     2.          1.         -7.
    -1.         -1.          1.
     1.         -1.          2.
```

Command: **transpose**
Transpose Matrix1 and store the transpose in Matrix2
Matrix1: a
Matrix2: <%> ap
Transpose A and store the transpose in AP

```
     5.          2.         -1.          1.
     0.          1.         -1.         -1.
     3.         -7.          1.          2.
```

```
Command: multiply ap a
Multiply AP by A and store in %
    31.          2.          2.
     2.          3.        -10.
     2.        -10.         63.
```

Note that $A'A$ is a symmetric matrix, as is always true.

The single command **compute** can be used to calculate an algebraic expression involving matrices.

Example 3: Let U be the column vector $\begin{pmatrix} 1/3 \\ -2/3 \\ 0 \\ 2/3 \end{pmatrix}$ and compute the matrix $(I - 2UU')^2$, where I is the identity matrix of appropriate size.

Here we first construct the matrix U. This is a two-step process: Construct a 4 by 1 matrix of zeros and then change its entries to the desired numbers.

```
Command: zero

Zero -- store the Integer1 by Integer2 zero matrix in Matrix1
Integer1: 4
Integer2: 1
Matrix1: <%> u
Zero -- store the 4 by 1 zero matrix in U
     0.
     0.
     0.
     0.
Command: change

Change the entries in Matrix1
Matrix1: u
Change the entries in U
U( 1, 1): <       0.  > 1/3
U( 2, 1): <       0.  > -2/3
U( 3, 1): <       0.  >
U( 4, 1): <       0.  > 2/3
```

Page 12

```
If you are done making changes, type EXIT
U( 4, 1): <      0.67> exit

     0.33
    -0.67
     0.
     0.67
```

Command: **compute**

```
Compute Expression1 and store in Matrix1
Expression1: (i-2*u*u')^2
Matrix1: <%> m
Compute ((I - ((2*U)*U'))^2) and store in M
      1.    -5.96E-08      0.          0.
  -5.96E-08       1.       0.      -1.34E-07
      0.         0.        1.          0.
      0.     -1.34E-07     0.          1.
```

Note that within the compute command I is interpreted as the identity
matrix of appropriate size to make the algebraic expression meaningful. The
entries in M not on the main diagonal are essentially 0. Hence the matrix
computed in this example is the 4 by 4 identity matrix.

PROBLEMS

IMPORTANT NOTE: For all problems in this book, the numbers in
brackets immediately following a problem number (or letter) denote matrices
in group storage that are to be used in that problem (or part).

1. [4 & 5, 6 & 7] For each of the indicated pairs of matrices A & B
   in group storage, compute the following matrices. Use the commands:
   get, add, subtract, multiply, transpose.

   a) $A + B$

   b) $4A$

   c) $4A - 7B$

   d) $-4AB'$

2. Redo Problem 1 using the **compute** command.

3. [8, 9, 10] For the indicated matrices A, B, C in group storage, do the following parts.

    a) Verify that $(AB)C = A(BC)$.

    b) Verify that $A(B + C) = AB + AC$.

    c) Verify that $AB \neq BA$.

    d) Verify that $(AB)' = B'A'$.

4. [11] Let A be the indicated matrix in group storage and let

$$E_1 = \begin{pmatrix} 1 & 0 & 0 \\ -3 & 1 & 0 \\ 0 & 0 & 1 \end{pmatrix}, \; E_2 = \begin{pmatrix} 1 & 0 & 0 \\ 0 & 0 & 1 \\ 0 & 1 & 0 \end{pmatrix}.$$

    a) Use the **zero** or **identity** commands followed by **change** or **edit** to construct E_1 and E_2.

    b) Compute $E_1 A$ and AE_1. What is the relationship between A and $E_1 A$? between A and AE_1?

    c) Compute $E_2 A$ and AE_2. What is the relationship between A and $E_2 A$? between A and AE_2?

5. For the indicated pair of matrices A & B, solve the given equation for the matrix X.

    a) [4 & 5] $2X + 4A = AB - 2X + A$

    b) [12 & 13] $10A'B + 3X = 6I + 8X + 12B'A$
       (Here I is an identity matrix.)

6. [9] A square matrix A is *idempotent* if $A^2 = A$. Verify that the indicated matrix is idempotent.

7. [14, 10, 15] A square matrix A is *nilpotent of index n $(n \geq 2)$* if $A^n = 0$ but $A^{n-1} \neq 0$. Verify that each indicated matrix is nilpotent and find its index.

8. [16] Verify that the indicated matrix A satisfies the polynomial equation $A^3 - 12A + 16I = 0$.

Section 1.2 NONSINGULAR MATRICES

An n by n matrix A is *nonsingular* or *invertible* if there exists an n by n matrix B such that $AB = BA = I$. If such a B exists, it is unique and is the *inverse* of A, denoted A^{-1}.

A key to determining the invertibility of a matrix is its *rank*, discussed in the next paragraph. The rank of A also determines the nature of the solutions of a matrix equation $AX = K$, as will be seen in Sections 1.3 and 1.4.

We can apply elementary row operations to a matrix A and reduce it to a matrix U in row-echelon form (i.e., the first nonzero entry in any row appears in a column to the right of the first nonzero entry in the preceding row). The procedure is called *Gaussian elimination* and U is a *reduced form* of A. The *rank* of A is the number of nonzero rows in U.

For example,

$$
A = \begin{pmatrix} 1 & -2 & 1 \\ 3 & -5 & 0 \\ -2 & 6 & 7 \end{pmatrix} \xrightarrow[\substack{-3R_1+R_2 \to R_2 \\ 2R_1+R_3 \to R_3}]{} \begin{pmatrix} 1 & -2 & 1 \\ 0 & 1 & -3 \\ 0 & 2 & 9 \end{pmatrix}
$$

$$
\xrightarrow[-2R_2+R_3 \to R_3]{} \begin{pmatrix} 1 & -2 & 1 \\ 0 & 1 & -3 \\ 0 & 0 & 15 \end{pmatrix} = U.
$$

(Here $-3R_1 + R_2 \to R_2$ means: Multiply the first row of A by -3 and add to the second row and replace the second row of A by the sum. Similarly for the other notations.) Since U has 3 nonzero rows, the rank of A is 3.

The relationship between invertibility and rank is as follows: An n by n matrix A is invertible iff the rank of A is n. (A proof is not so direct; see any linear algebra text.)

Example 1: Use MAX to record the rank of $A = \begin{pmatrix} 1 & -2 & 1 \\ 3 & -5 & 0 \\ -2 & 6 & 7 \end{pmatrix}$.

```
Command: get 17 a
Get matrix 17 and store it in A
        1.        -2.        1.
        3.        -5.        0.
       -2.         6.        7.
```

```
Command: rank
Rank -- record the rank of Matrix1
Matrix1: a
Rank -- record the rank of A
The following upper-triangular matrix has the same rank as A:
     8.06         4.96         -3.6
                 -5.04         -0.96
                               -0.37
Determinant =   1.50E1      r-condition =   2.30E-02
Enter the rank of A (it is at most 3): 3
```

Some explanation of this output is in order. The upper-triangular matrix computed by MAX is not the U we computed above. The program employs somewhat complicated algorithms to ensure greater numerical accuracy. (See Appendix II for more information.) Since A is square, MAX also computes its determinant and r-condition (reciprocal condition number — again see Appendix II). All that we need be concerned with is the following: The upper-triangular matrix displayed on the screen has the same number of nonzero rows as a reduced form of A we might compute by hand. Hence we can use it to record the rank of A. Round-off error sometimes makes it difficult to judge if a number is zero. Consequently, the program refrains from supplying the rank of a matrix; we must enter it ourselves. The present matrix has rank 3 so we enter 3 in response to the prompt.

Having convinced ourselves that A is invertible, we find its inverse.

```
Command: invert a
Invert A and store the inverse in %
    -2.33         1.33         0.33
    -1.4          0.6          0.2
     0.53        -0.13      6.67E-02
```

This is the matrix A^{-1}. If we wish further confirmation, we can multiply A and A^{-1}.

```
Command: compute a^-1*a
Compute ((A^-1)*A) and store in %
        1.          0.          0.
  1.49E-07          1.          0.
        0.          0.          1.
Command: compute a*a^-1
Compute (A*(A^-1)) and store in %
        1.          0.          0.
 -4.77E-07          1.    1.19E-07
 -4.77E-07    1.19E-07          1.
Command: round %
Round the entries of % to integers and store the result in %
        1.          0.          0.
        0.          1.          0.
        0.          0.          1.
```

Example 2: Let B be matrix 4 in group storage. Find the rank of B and determine whether B is nonsingular.

```
Command: get 4 b
Get matrix 4 and store it in B
        1.          2.          3.
        3.          2.          1.
        0.          2.          4.
Command: rank b
Rank -- record the rank of B

The following upper-triangular matrix has the same rank as B:
     -5.1       -1.18       -3.14
                -2.94       -1.47
                             2.38E-07
Determinant =  0.00E0     r-condition =        0.
Enter the rank of B (it is at most 3): 2
```

From the information, we conclude that B has rank 2 and hence is a singular matrix. (You might see what happens if you try to invert B.)

PROBLEMS

1. [18, 19, 20] For each of the indicated n by n matrices A, find the rank of A. If A has rank n, find A^{-1}.

2. [21 & 1, 22 & 23] For each of the indicated pairs of matrices P & A, check that P is nonsingular, compute $P^{-1}AP$, and confirm that this product is a diagonal matrix.

3. [8, 20] For each of the indicated matrices, do the following parts.

   a) Find A^2 and $(A^2)^{-1}$. Confirm that $(A^2)^{-1} = (A^{-1})^2$.
   Is $(A^3)^{-1} = (A^{-1})^3$?

   b) Find $(A')^{-1}$ and confirm that $(A')^{-1} = (A^{-1})'$.

   c) Find $(10A)^{-1}$ and confirm that $(10A)^{-1} = \frac{1}{10}A^{-1}$.

4. [17 & 18, 20 & 35] For each of the indicated pairs of matrices A & B, find $(AB)^{-1}$ and confirm that $(AB)^{-1} = B^{-1}A^{-1}$.

5 Let A and B be nonsingular matrices.

   a) Is AB nonsingular?

   b) Is $A + B$ nonsingular? (Try some examples.)

6. Let J be the 4 by 4 matrix each of whose entries is 1 and let I be the 4 by 4 identity matrix.

   a) Verify that $(I - J)^{-1} = I - \frac{1}{3}J$.
   (Use the **ones** command to construct J.)

   b) What similar equality holds for n by n matrices?

   c) Prove your assertion in part b).

7. a) [24] For the indicated matrix A, show that $A^2 - 5A + 4I = 0$. Use this equation to express A^{-1} as a polynomial in A.
   Compute A^{-1} from this polynomial.
   Check using the **invert** command.

   b) [25] For the indicated matrix A, show that $A^3 - A^2 - A + I = 0$.
Use this equation to express A^{-1} as a polynomial in A.
Compute A^{-1} from this polynomial.
Check using the **invert** command.

8. [14, 10, 15] Recall (Problem 7 in Section 1.1) that the indicated matrices A are nilpotent. Do the following parts for each matrix A.

   a) Use the fact that $A^n = 0$ to express the inverse of $I - A$ as a polynomial. Then use MAX to compute $(I - A)^{-1}$ from this polynomial.

   b) Check your answer in part a) by computing $(I - A)^{-1}$ using the **invert** command.

Section 1.3 THE SYSTEM $AX = K$, A NONSINGULAR

Consider the system of equations

$$\begin{array}{rrrrr} x_1 & - & 2x_2 & + & x_3 & = & 3 \\ 3x_1 & - & 5x_2 & & & = & -5 \\ -2x_1 & + & 6x_2 & + & 7x_3 & = & -4. \end{array}$$

We solve this system by hand by writing it in matrix form and reducing:

$$\left(\begin{array}{rrr|r} 1 & -2 & 1 & 3 \\ 3 & -5 & 0 & -5 \\ -2 & 6 & 7 & -4 \end{array} \right) \xrightarrow[\substack{-3R_1+R_2 \to R_2 \\ 2R_1+R_3 \to R_3}]{} \left(\begin{array}{rrr|r} 1 & -2 & 1 & 3 \\ 0 & 1 & -3 & -14 \\ 0 & 2 & 9 & 2 \end{array} \right)$$

$$\xrightarrow[-2R_2+R_3 \to R_3]{} \left(\begin{array}{rrr|r} 1 & -2 & 1 & 3 \\ 0 & 1 & -3 & -14 \\ 0 & 0 & 15 & 30 \end{array} \right).$$

Solve this last system by back-substitution: $15x_3 = 30$, $x_3 = 2$, $x_2 = 3x_3 - 14 = 6 - 14 = -8$, $x_1 = 2x_2 - x_3 + 3 = -16 - 2 + 3 = -15$.

The given system has the unique solution

$$\begin{aligned} x_1 &= -15 \\ x_2 &= -8 \\ x_3 &= 2. \end{aligned}$$

This example generalizes. Consider a system of n equations in n unknowns written in matrix form as $AX = K$. Assume A is nonsingular, i.e., the rank of A is n. If we reduce the augmented matrix $\left(A \; K \right)$ to $\left(U \; H \right)$, U in row-echelon form, then $\left(U \; H \right)$ has the form

$$\left(\begin{array}{ccccc|c} * & & & & & \\ & * & & & & \\ & & \ddots & & & \\ & & & \ddots & & \\ & & & & * & \\ & O & & & & \end{array} \right)$$

where the $*$'s indicate nonzero numbers. This system is solved by back-substitution (first find x_n, then $x_{n-1}, \ldots$, then x_1). Hence $AX = K$ has a unique solution whenever A is nonsingular.

Example 1: Use MAX to find the unique solution of the above system of equations $AX = K$.

```
Command: get 17 a
Get matrix 17 and store it in A

        1.          -2.          1.
        3.          -5.          0.
       -2.           6.          7.

Command: get 26 k
Get matrix 26 and store it in K

        3.
       -5.
       -4.

Command: solve

Solve the system Matrix1 Matrix2 and store the solution in Matrix3
Matrix1: a
Matrix2: k
Matrix3: <%>

Enter the rank of A (it is at most 3): 3
Solve the system A K and store the solution in %

      -15.
       -8.
        2.
```

Note that the rank of the coefficient matrix A must be provided before the system $AX = K$ can be solved.

Example 2: Let A and K be matrices 27 and 28 in group storage. Solve the equation $3X = AX + K$ for X.

Rewrite the equation as $3X - AX = K$ or $(3I - A)X = K$ and use the **solve** command.

```
Command: get 27 a
Get matrix 27 and store it in A

        1.           1.          1.          0.
        2.          -2.          0.          2.
        4.           4.         -4.          0.
        8.          -8.          0.         -8.
```

```
Command: get 28 k
Get matrix 28 and store it in K
    14.
   -41.
    -1.
   -61.
Command: compute 3*i - a b
Compute ((3*I) - A) and store in B
     2.         -1.         -1.          0.
    -2.          5.          0.         -2.
    -4.         -4.          7.          0.
    -8.          8.          0.         11.
Command: rank b
Rank -- record the rank of B
The following upper-triangular matrix has the same rank as B:
   -11.18      -6.98       0.          7.51
               -7.57       3.57        1.
                          -6.11        5.5
                                      -0.56
Determinant =   2.92E2      r-condition =   1.62E-02
Enter the rank of B (it is at most 4): 4
Command: solve b k
Solve the linear system B K and store the solution in %
     2.
    -7.
    -3.
     1.
```

Note: Whenever A is nonsingular, MAX uses the LU factorization of A to solve a system $AX = K$. LU factorization can be explained as follows. Recall again the example used in this section and the preceding:

$$A = \begin{pmatrix} 1 & -2 & 1 \\ 3 & -5 & 0 \\ -2 & 6 & 7 \end{pmatrix} \xrightarrow[\substack{-3R_1+R_2\to R_2 \\ 2R_1+R_3\to R_3}]{} \begin{pmatrix} 1 & -2 & 1 \\ 0 & 1 & -3 \\ 0 & 2 & 9 \end{pmatrix}$$

$$\xrightarrow[-2R_2+R_3\to R_3]{} \begin{pmatrix} 1 & -2 & 1 \\ 0 & 1 & -3 \\ 0 & 0 & 15 \end{pmatrix} = U.$$

Performing these elementary row operations to reduce A to U amounts to multiplying U on the left by a lower-triangular matrix L with 1's on the diagonal and the negatives of the multipliers -3, 2, and -2 below the diagonal. That is,

$$A = \begin{pmatrix} 1 & -2 & 1 \\ 3 & -5 & 0 \\ -2 & 6 & 7 \end{pmatrix} = \begin{pmatrix} 1 & 0 & 0 \\ 3 & 1 & 0 \\ -2 & 2 & 1 \end{pmatrix} \begin{pmatrix} 1 & -2 & 1 \\ 0 & 1 & -3 \\ 0 & 0 & 15 \end{pmatrix} = LU.$$

(Check to make sure.) Often, unlike this example, row interchanges are performed in reducing A to U. Interchanging rows of A amounts to multiplying A on the left by a permutation matrix P, i.e., a matrix that results from performing row interchanges on the identity matrix. When row interchanges occur, the LU factorization of A becomes $PA = LU$.

To summarize, LU factorization of A is a way of looking at the process of Gaussian elimination as a whole; elementary row operations applied to A amount to the LU factorization of A.

In computing the L and U factors of a matrix, MAX employs Gaussian elimination with *partial pivoting*: When introducing zeros below an entry on the diagonal, a row interchange is performed so that the diagonal entry (the pivot) is at least as large in absolute value as all the entries below it. (This is done to increase numerical accuracy.) In the above example, the steps are:

$$A = \begin{pmatrix} 1 & -2 & 1 \\ 3 & -5 & 0 \\ -2 & 6 & 7 \end{pmatrix} \xrightarrow[R_1 \leftrightarrow R_2]{} \begin{pmatrix} 3 & -5 & 0 \\ 1 & -2 & 1 \\ -2 & 6 & 7 \end{pmatrix} \xrightarrow[\substack{-\frac{1}{3}R_1 + R_2 \to R_2 \\ \frac{2}{3}R_1 + R_3 \to R_3}]{} \begin{pmatrix} 3 & -5 & 0 \\ 0 & -1/3 & 1 \\ 0 & 8/3 & 7 \end{pmatrix}$$

$$\xrightarrow[R_2 \leftrightarrow R_3]{} \begin{pmatrix} 3 & -5 & 0 \\ 0 & 8/3 & 7 \\ 0 & -1/3 & 1 \end{pmatrix} \xrightarrow[\frac{1}{8}R_2 + R_3 \to R_3]{} \begin{pmatrix} 3 & -5 & 0 \\ 0 & 8/3 & 7 \\ 0 & 0 & 15/8 \end{pmatrix}.$$

This LU factorization of A is

$$PA = \begin{pmatrix} 0 & 1 & 0 \\ 0 & 0 & 1 \\ 1 & 0 & 0 \end{pmatrix} \begin{pmatrix} 1 & -2 & 1 \\ 3 & -5 & 0 \\ -2 & 6 & 7 \end{pmatrix} =$$

$$\begin{pmatrix} 1 & 0 & 0 \\ -2/3 & 1 & 0 \\ 1/3 & -1/8 & 1 \end{pmatrix} \begin{pmatrix} 3 & -5 & 0 \\ 0 & 8/3 & 7 \\ 0 & 0 & 15/8 \end{pmatrix} = LU.$$

```
Command: get 17 a
Get matrix 17 and store it in A
        1.        -2.        1.
        3.        -5.        0.
       -2.         6.        7.
Command: lufactor

LUfactor Matrix1 and store the L and U factors in [Matrix2 Matrix3]
Matrix1: a
Matrix2: <%> l
Matrix3: <%%> u
LUfactor A and store the L and U factors in [L U]

The product of the following L and U factors is A with row swaps:
   1 <--> 2, 2 <--> 3
        1.         0.         0.
       -0.67       1.         0.
        0.33      -0.12       1.

        3.        -5.         0.
        0.         2.67       7.
        0.         0.         1.87
```

The LU factorization of A is used to solve a system $AX = K$ as follows. From $PA = LU$, we have $LUX = PK$. Solve the lower-triangular system $L(UX) = PK$ for UX by forward substitution; denote the solution by C. Then solve the upper-triangular system $UX = C$ for X by back-substitution. Symbolically, we have

$$(P^{-1}L)(UX) = K,$$
$$UX = (P^{-1}L)^{-1}K,$$
$$X = U^{-1}(P^{-1}L)^{-1}K$$

(though the inverse matrices are not actually computed).

PROBLEMS

1. [18 & 29, 19 & 30, 20 & 44] For each of the indicated pairs of matrices A & K, do the following parts.

   a) Find the rank of A. Is A nonsingular? Why?

   b) If A is nonsingular, find the unique solution of $AX = K$. Use the **solve** command.

   c) If A is nonsingular, solve $AX = K$ again, this time using A^{-1} and the **compute** command.

2. [31] For the indicated matrix A, consider the system of equations $AX = 0$.

   a) Find the rank of A.

   b) Construct the 4 by 1 zero matrix 0. Solve the system $AX = 0$.

   c) Complete the following sentence: If A is an n by n matrix, $AX = 0$ has the unique solution $X =$ _____ iff rank $A =$ _____. Prove your assertion.

3. For each of the indicated pairs of matrices A & K, solve the given matrix equation for X.

   a) [31 & 32] $AX + 5K + 4X = 0$

   b) [33 & 34] $5AX + K = A'X - 2X + 9K$

4. [35; 36, 37, 38, 39] For the indicated matrix A and the four indicated matrices K, solve the equation $AX = K$. Find all four solutions with a single application of the **solve** command.

5. [40, 41] For each of the indicated matrices A, find A^{-1} two ways:

   a) using the **invert** command,

   b) using the **solve** command.

6. [18, 42, 20] For each of the indicated matrices A, find the LU factorization of A. Find the permutation matrix P such that $PA = LU$.

7. [18 & 29, 42 & 43, 20 & 44] For each of the indicated pairs of matrices A & K, use the LU factorization of A to find the unique solution of $AX = K$. Compare your answers with those obtained in Problem 1.

Section 1.4 THE SYSTEM $AX = K$, A ARBITRARY

Suppose we solve the system of equations

$$\begin{array}{rrrrrrr}
x_1 & + & 3x_2 & + & 2x_3 & + & x_4 & - & x_5 & = & 7 \\
7x_1 & - & 3x_2 & + & 2x_3 & + & x_4 & - & 7x_5 & = & 1 \\
3x_1 & + & x_2 & + & 2x_3 & + & x_4 & - & 3x_5 & = & 5
\end{array}$$

by hand using Gaussian elimination and back-substitution. We first write in matrix form and reduce.

$$\begin{pmatrix}
1 & 3 & 2 & 1 & -1 & | & 7 \\
7 & -3 & 2 & 1 & -7 & | & 1 \\
3 & 1 & 2 & 1 & -3 & | & 5
\end{pmatrix}$$

$$\underset{\substack{-7R_2+R_2\to R_2 \\ -3R_1+R_3\to R_3}}{\longrightarrow}
\begin{pmatrix}
1 & 3 & 2 & 1 & -1 & | & 7 \\
0 & -24 & -12 & -6 & 0 & | & -48 \\
0 & -8 & -4 & -2 & 0 & | & -16
\end{pmatrix}$$

$$\underset{-\frac{1}{3}R_2+R_3\to R_3}{\longrightarrow}
\begin{pmatrix}
1 & 3 & 2 & 1 & -1 & | & 7 \\
0 & -24 & -12 & -6 & 0 & | & -48 \\
0 & 0 & 0 & 0 & 0 & | & 0
\end{pmatrix}$$

$$\underset{-\frac{1}{6}R_2\to R_2}{\longrightarrow}
\begin{pmatrix}
1 & 3 & 2 & 1 & -1 & | & 7 \\
0 & 4 & 2 & 1 & 0 & | & 8 \\
0 & 0 & 0 & 0 & 0 & | & 0
\end{pmatrix}.$$

The corresponding system of equations is

$$\begin{array}{rrrrrrr}
x_1 & + & 3x_2 & + & 2x_3 & + & x_4 & - & x_5 & = & 7 \\
& & 4x_2 & + & 2x_3 & + & x_4 & & & = & 8 \,.
\end{array}$$

Here x_3, x_4, x_5 can be chosen arbitrarily and we solve for x_1, x_2 by back-substitution. Let $x_3 = c_1, x_4 = c_2, x_5 = c_3$. Then

$$4x_2 = -2x_3 - x_4 + 8 = -2c_1 - c_2 + 8, \quad x_2 = -\frac{1}{2}c_1 - \frac{1}{4}c_2 + 2,$$

$$x_1 = -3x_2 - 2x_3 - x_4 + x_5 + 7 = -3\left(-\frac{1}{2}c_1 - \frac{1}{4}c_2 + 2\right) - 2c_1 - c_2 + c_3 + 7$$

$$= -\frac{1}{2}c_1 - \frac{1}{4}c_2 + c_3 + 1.$$

The general solution of $AX = K$ is

$$X = \begin{pmatrix} x_1 \\ x_2 \\ x_3 \\ x_4 \\ x_5 \end{pmatrix} = \begin{pmatrix}
-1/2c_1 & -1/4c_2 & + & c_3 & + & 1 \\
-1/2c_1 & -1/4c_2 & & & + & 2 \\
c_1 & & & & & \\
& c_2 & & & & \\
& & & c_3 & &
\end{pmatrix}$$

$$= c_1 \begin{pmatrix} -1/2 \\ -1/2 \\ 1 \\ 0 \\ 0 \end{pmatrix} + c_2 \begin{pmatrix} -1/4 \\ -1/4 \\ 0 \\ 1 \\ 0 \end{pmatrix} + c_3 \begin{pmatrix} 1 \\ 0 \\ 0 \\ 0 \\ 1 \end{pmatrix} + \begin{pmatrix} 1 \\ 2 \\ 0 \\ 0 \\ 0 \end{pmatrix}.$$

Suppose we write the solution as $X = c_1 X_1 + c_2 X_2 + c_3 X_3 + X_p$. Then X_1, X_2, X_3 are solutions of the homogeneous equation $AX = 0$, and X_p is a particular solution of the given equation $AX = K$ (obtained by setting $c_1 = c_2 = c_3 = 0$).

Now consider the system of equations

$$\begin{aligned} x_1 + 3x_2 + 2x_3 + x_4 - x_5 &= 7 \\ 7x_1 - 3x_2 + 2x_3 + x_4 - 7x_5 &= 1 \\ 3x_1 + x_2 + 2x_3 + x_4 - 3x_5 &= 4. \end{aligned}$$

The matrix reduction now becomes

$$\begin{pmatrix} 1 & 3 & 2 & 1 & -1 & | & 7 \\ 7 & -3 & 2 & 1 & -7 & | & 1 \\ 3 & 1 & 2 & 1 & -3 & | & 4 \end{pmatrix} \rightarrow \begin{pmatrix} 1 & 3 & 2 & 1 & -1 & | & 7 \\ 0 & -24 & -12 & -6 & 0 & | & -48 \\ 0 & -8 & -4 & -2 & 0 & | & -17 \end{pmatrix}$$

$$\rightarrow \begin{pmatrix} 1 & 3 & 2 & 1 & -1 & | & 7 \\ 0 & -24 & -12 & -6 & 0 & | & -48 \\ 0 & 0 & 0 & 0 & 0 & | & -1 \end{pmatrix}.$$

The third equation of the corresponding system is $0x_1 + 0x_2 + 0x_3 + 0x_4 + 0x_5 = -1$. Since there are no solutions to this equation, the given system has no solutions.

These examples generalize to give the following results. Let A be an m by n matrix of rank r, let K be an m by 1 matrix, and consider the system $AX = K$. Then

If $r = \text{rank } A < \text{rank} \begin{pmatrix} A & K \end{pmatrix}$, $AX = K$ has no solutions.

If $r = \text{rank } A = \text{rank} \begin{pmatrix} A & K \end{pmatrix}$, $AX = K$ has a solution.

If $r = n$, there is a unique solution X_p.

If $r < n$, there are infinitely many solutions, each of the form

$c_1 X_1 + \ldots + c_{n-r} X_{n-r} + X_p$ where

X_p is a particular solution of $AX = K$,

$X_1, \ldots, X_{n-r}$ are linearly independent solutions of $AX = 0$,

$c_1, \ldots, c_{n-r}$ are arbitrary constants.

In fact, $X_1, \ldots, X_{n-r}$ form a basis for the nullspace of A (the set of all solutions of $AX = 0$), and the nullspace has dimension $n - r$. This vector space terminology is discussed in Chapter 2; it is used in the MAX commands that are employed to solve systems of equations.

Example 1: Use MAX to solve the system $AX = K$ where

$$A = \begin{pmatrix} 1 & 3 & 2 & 1 & -1 \\ 7 & -3 & 2 & 1 & -7 \\ 3 & 1 & 2 & 1 & -3 \end{pmatrix}, K = \begin{pmatrix} 7 \\ 1 \\ 5 \end{pmatrix}.$$

```
Command: get 45 a
Get matrix 45 and store it in A
     1.         3.         2.         1.        -1.
     7.        -3.         2.         1.        -7.
     3.         1.         2.         1.        -3.
Command: get 46 k
Get matrix 46 and store it in K
     7.
     1.
     5.
Command: rank a
Rank -- record the rank of A

The following upper-triangular matrix has the same rank as A:
    -7.68        1.95       -2.86       -1.43        7.68
                 3.9         1.95        0.97        0.
                            -1.79E-07  -8.94E-08     0.

Enter the rank of A (it is at most 3): 2

Command: augment

Augment Matrix1 by Matrix2 and store the result in Matrix3
Matrix1: a
Matrix2: k
Matrix3: <%> ak
Augment A by K and store the result in AK
     1.         3.         2.         1.        -1.        7.
     7.        -3.         2.         1.        -7.        1.
     3.         1.         2.         1.        -3.        5.
```

Command: **rank ak**

Rank -- record the rank of AK

The following upper-triangular matrix has the same rank as AK:

```
    -8.66       3.35      -2.66      -1.5       -3.        -3.35
                6.91       3.46      -0.86      -1.73      -6.91
                           5.96E-07  1.04E-07  2.09E-07  -1.19E-07
```

Enter the rank of AK (it is at most 3): **2**

Since rank $A = \text{rank} \left(A\ K \right)$, the system has solutions. We find a particular solution X_p using the **solve** command.

Command: **solve a k xp**

Solve the system A K and store the solution in XP

```
    1.
    2.
    0.
    0.
    0.
```

WARNING: This is a least-squares solution, which need not be an
         actual solution

Having checked ranks, we can ignore the warning since we know an actual solution exists. We find a basis for the nullspace of A using the **nbasis** command.

Command: **nbasis**

NBasis -- find a nullspace basis of Matrix1 and store in Matrix2
Matrix1: a
Matrix2: <%> xh
NBasis -- find a nullspace basis of A and store in XH

```
    -0.5      -0.25      1.
    -0.5      -0.25      0.
     1.        0.        0.
     0.        1.        0.
     0.        0.        1.
```

The columns of this matrix are the basis vectors X_1, X_2, X_3 and the general solution of $AX = K$ is $X = c_1 X_1 + c_2 X_2 + c_3 X_3 + X_p$.

Note that this solution is identical to the one obtained above; indeed, the example was chosen to work out this way. Since the X_i and X_p are not unique, often the general solution from MAX will not look the same as that computed by hand. See Example 3 below.

Example 2: Use MAX to show that $AX = K$ has no solutions where

$$A = \begin{pmatrix} 1 & 3 & 2 & 1 & -1 \\ 7 & -3 & 2 & 1 & -7 \\ 3 & 1 & 2 & 1 & -3 \end{pmatrix}, K = \begin{pmatrix} 7 \\ 1 \\ 4 \end{pmatrix}.$$

Suppose the matrices A, K and AK from the previous example are still in the workspace. We know rank $A = 2$.

```
Command: change ak(3,6)
Change the entries in AK(3-3,6-6)
AK( 3, 6): <      5. > 4
     1.        3.        2.        1.       -1.        7.
     7.       -3.        2.        1.       -7.        1.
     3.        1.        2.        1.       -3.        4.
Command: rank ak
Rank -- record the rank of AK

The following upper-triangular matrix has the same rank as AK:
    -8.12       3.2      -2.71      -1.48      -2.95      -3.2
                6.98      3.39      -0.9       -1.8       -6.98
                          0.42      0.11       0.21       0.
Enter the rank of AK (it is at most 3): 3
```

Since rank $A < $ rank $\begin{pmatrix} A & K \end{pmatrix}$, the system has no solutions.

Example 3: Solve $AX = K$ where A is matrix 4 and K is matrix 47.

```
Command: get 4 a
Get matrix 4 and store it in A
     1.        2.        3.
     3.        2.        1.
     0.        2.        4.
```

```
Command: get 47 k
Get matrix 47 and store it in K

     -1.
      1.
     -2.

Command: rank a
Rank -- record the rank of A

The following upper-triangular matrix has the same rank as A:

     -5.1        -1.18       -3.14
                 -2.94       -1.47
                              2.38E-07

Determinant =  0.00E0     r-condition =      0.

Enter the rank of A (it is at most 3): 2

Command: aug a k ak
Augment A by K and store the result in AK

      1.          2.          3.         -1.
      3.          2.          1.          1.
      0.          2.          4.         -2.

Command: rank ak
Rank -- record the rank of AK

The following upper-triangular matrix has the same rank as AK:

     -5.1        -1.18       -3.14        1.96
                 -2.94       -1.47       -1.47
                              2.38E-07    0.

Enter the rank of AK (it is at most 3): 2

Command: solve a k xp
Solve the system A K and store the solution in XP

      0.5
      0.
     -0.5

WARNING: This is a least-squares solution, which need not be an
         actual solution
```

```
Command: nbasis a xh
NBasis -- find a nullspace basis of A and store in XH
    -0.5
     1.
    -0.5
```

The general solution of $AX = K$ is $X = c_1 \begin{pmatrix} -1/2 \\ 1 \\ -1/2 \end{pmatrix} + \begin{pmatrix} 1/2 \\ 0 \\ -1/2 \end{pmatrix}$.

For comparison, also solve this system by hand and show that the solution you get is equivalent to this one, even though the solutions look different.

Remark: As noted at the end of the preceding section, the **solve** command in MAX uses the LU factorization of A to find the unique solution of $AX = K$ when A is nonsingular. If A is singular, the **solve** command in MAX uses the QR factorization of A to find a least-squares solution of $AX = K$. This is explained in Section 2.6, with additional details in Appendix II.

PROBLEMS

1. For this first problem, recall that the nullspace of A is the set of solutions of $AX = 0$.

   [48, 49, 50] For each of the indicated m by n matrices A, do the following parts.

   a) Find the rank of A.

   b) Find a basis for the nullspace of A.

   c) Pick any two solutions $X1, X2$ of $AX = 0$ and pick any two constants a, b. Check that $aX1 + bX2$ is a solution of $AX = 0$.

   d) Without further computation, give all solutions of $AX = 0$.

   e) Complete the following sentences:
   $AX = 0$ has a unique solution iff rank A _____.
   $AX = 0$ has a nonzero solution iff rank A _____.

   f) Prove your assertions in part e).

   g) The dimension of the nullspace of A is called the *nullity* of A. Complete the following sentence: rank A + nullity A = _____.

   h) Prove your assertion in part g).

2. [31 & 51, 19 & 30, 12 & 52, 13 & 53] For each of the indicated pairs of matrices A & K, do the following parts.

   a) Find the rank of A and of $\begin{pmatrix} A & K \end{pmatrix}$.

   b) Does $AX = K$ have a solution? Why?

   c) Find a solution of $AX = K$, if there is one.

   d) Find all solutions of $AX = 0$.

   e) Pick any solution $X1$ of $AX = 0$ and any solution $Y1$ of $AX = K$. Check that $X1 + Y1$ is a solution of $AX = K$.

   f) Without further computation, give all solutions of $AX = K$.

3. [54 & 55, 56 & 57, 58 & 32] For each of the indicated pairs of matrices A & K, find all solutions of the system $AX = K$.

4. In several hand computations in the past two sections, we used elementary row operations to reduce a matrix A to a matrix U in row-echelon form. For example,

$$A = \begin{pmatrix} 1 & 1 & 2 & -1 \\ 2 & 4 & -3 & 1 \\ 3 & -1 & -5 & 6 \end{pmatrix} \xrightarrow[\substack{-2R_1+R_2 \to R_2 \\ -3R_1+R_3 \to R_3}]{} \begin{pmatrix} 1 & 1 & 2 & -1 \\ 0 & 2 & -7 & 3 \\ 0 & -4 & -11 & 9 \end{pmatrix}$$

$$\xrightarrow[2R_2+R_3 \to R_3]{} \begin{pmatrix} 1 & 1 & 2 & -1 \\ 0 & 2 & -7 & 3 \\ 0 & 0 & -25 & 15 \end{pmatrix} = U.$$

If we wish, we can use MAX to carry out this reduction one step at a time.

```
Command: get 59 a
Get matrix 59 and store it in A
     1.        1.        2.       -1.
     2.        4.       -3.        1.
     3.       -1.       -5.        6.
```

```
Command: compute -2*a(1,) + a(2,) a(2,)
Compute ((-2*A(1-1,1-4)) + A(2-2,1-4)) and store in A(2-2,1-4)

     1.        1.        2.        -1.
     0.        2.        -7.        3.
     3.       -1.        -5.        6.

Command: compute -3*a(1,) + a(3,) a(3,)
Compute ((-3*A(1-1,1-4)) + A(3-3,1-4)) and store in A(3-3,1-4)

     1.        1.        2.        -1.
     0.        2.        -7.        3.
     0.       -4.       -11.        9.

Command: compute 2*a(2,) + a(3,) a(3,)
Compute ((2*A(2-2,1-4)) + A(3-3,1-4)) and store in A(3-3,1-4)

     1.        1.        2.        -1.
     0.        2.        -7.        3.
     0.        0.       -25.       15.
```

---

    a) [60, 61] For each of the indicated matrices A, reduce A one step at a time to a matrix U in row-echelon form. (No row interchanges are needed.)

    b) [62, 63] For each of the indicated matrices A, reduce A one step at a time to a matrix U in row-echelon form. (Use the **swap** command to interchange rows when necessary.)

5. [2 & 65, 63 & 64] For each of the indicated pairs of matrices A & K, solve the system $AX = K$ by reducing the augmented matrix $\left(A\ K \right)$ one step at a time to a matrix in row-echelon form and using back-substitution.

6. [66, 67] For each of the indicated matrices A, reduce A one step at a time to a matrix U in row-echelon form using partial pivoting (recall Section 1.3). Check that your answer is the same as the U factor in the LU factorization of A.

Section 1.5 APPLICATIONS

In this section, we describe situations that give rise to systems of linear equations, written in matrix form as $AX = K$. In many cases, the matrices are large enough to make hand calculation tedious at best.

PROBLEMS

Temperature Distributions

Consider a thin plate whose temperature along the edge is given. We wish to find the equilibrium temperature at points inside the plate. To approximate the temperature, we consider a rectangular grid of points in the plate and make the following assumption:

> At each grid point in the interior of the plate, the temperature is the average of the temperatures at the four surrounding grid points.

For example, consider the I-beam shown supporting a bridge roadway.

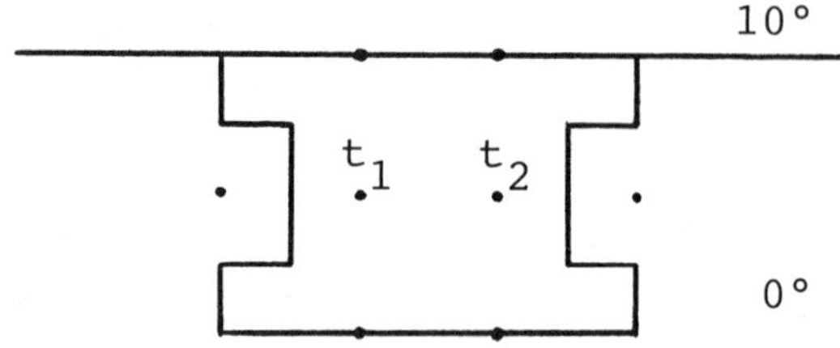

The roadway has a temperature of $10°$ and the air temperature is $0°$. The temperatures at the two pictured grid points satisfy the system of equations:

$$t_1 = \frac{1}{4}(0 + t_2 + 0 + 10)$$

$$t_2 = \frac{1}{4}(t_1 + 0 + 0 + 10)$$

or $T = AT + B$ where $T = \begin{pmatrix} t_1 \\ t_2 \end{pmatrix}, A = \begin{pmatrix} 0 & 1/4 \\ 1/4 & 0 \end{pmatrix}, B = \begin{pmatrix} 10/4 \\ 10/4 \end{pmatrix}$. If we solve this system, we get $t_1 = t_2 = \frac{10}{3}$.

In general, n interior grid points give rise to n equations in n unknowns written in matrix form as $T = AT + B$. Such a matrix equation can be solved using MAX.

Note: In the next four figures, $t_1, \ldots, t_n$ will denote the temperatures at the indicated interior points numbered by rows from left to right.

1. [81 & 82] Suppose we place a finer grid on the I-beam as shown.

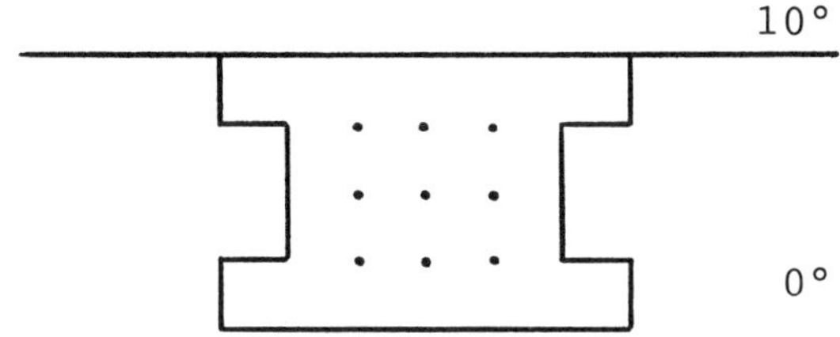

a) Derive the 9 by 9 matrix A and 9 by 1 matrix B so that
$T = AT + B$.
Compare your answers with matrices 81 and 82.

b) Find the temperatures at the 9 interior grid points.

2. [83 & 84] Suppose we redesign the I-beam as shown in the figure.

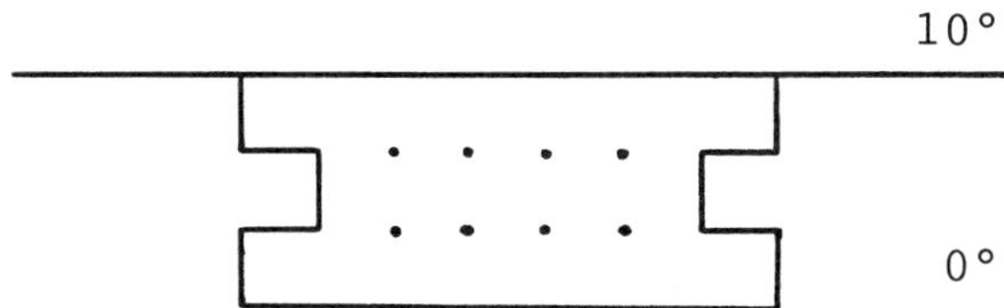

a) Derive the matrices A and B so that $T = AT + B$.
Compare your answers with matrices 83 and 84.

b) Find the temperatures at the interior grid points.

3. [85 & 86] Repeat Problem 2 for the I-beam as shown in the figure.

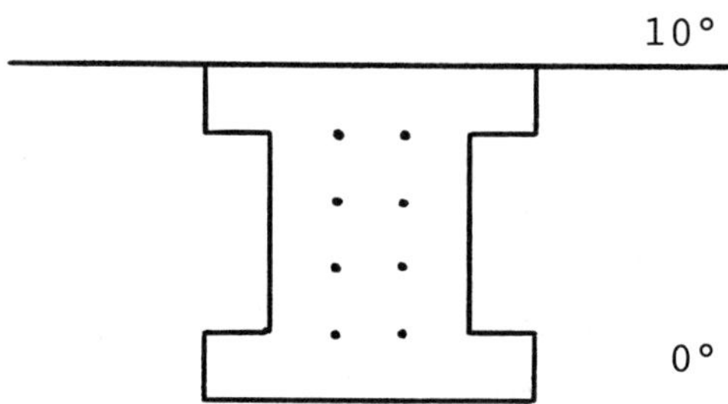

4. [87 & 82] Repeat Problem 2 for the beam as shown in the figure.

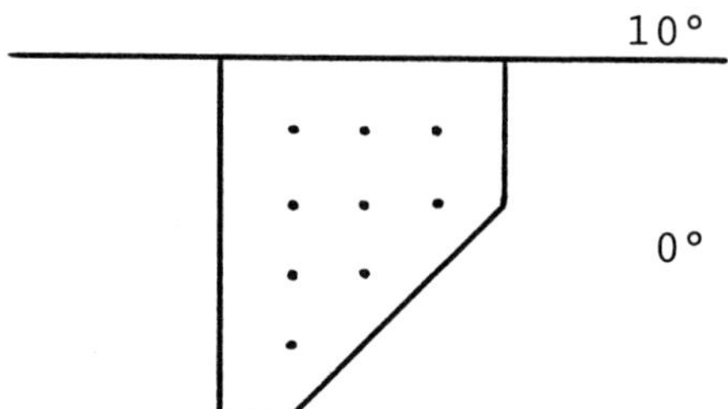

Leontief economic models

Consider an economy composed of n companies (or sectors), each producing one product, and an open sector that supplies no product but has a demand for the n products produced by the companies.

For i and j running from 1 to n, let

x_i = number of dollars' worth of product i supplied to all sectors,

d_i = number of dollars' worth of product i consumed by the open sector (the *final demand* for product i),

a_{ij} = number of dollars' worth of product i used to make one dollar's worth of product j (the *techological constants*).

The total demand for product i is

$$a_{i1}x_1 + \ldots + a_{in}x_n + d_i.$$

In equilibrium, supply equals demand and so

$$x_i = \sum_{j=1}^{n} a_{ij}x_j + d_i \quad \text{for each} \quad i = 1, \ldots, n.$$

In matrix notation, this system of equations becomes

$$X = AX + D \quad \text{where} \quad X = \left(x_i \right), A = \left(a_{ij} \right), D = \left(d_i \right).$$

If A and D are known, we can use MAX to find X, the vector of production levels (or *production vector*). Often one wishes to fix the technological constants A and calculate X for various demands D.

The situation described above is an *n-company Leontief model.*

5. [68 & 69] Consider an economy where agriculture, manufacturing, and transportation are the producing sectors. Suppose the flow of product, in millions of dollars, is given by the table below. (For instance, agriculture sells $75 million of its product to manufacturing.)

   The fourth column is the open sector, and the fifth column is the total value of the products produced. (For instance, agriculture sells $70 million of its products outside the three-sector economy.)

   The value-added entry in each column represents the value which that sector adds to the products it consumes. That is, it is the difference between the total value of products it produces and the products it consumes to support this production.

↓ From/To →	Ag.	Mfg.	Tr.	Final Demand	Total Produced
Agriculture	60	75	45	70	250
Manufacturing	100	60	63	152	375
Transportation	40	45	15	50	150
Total Consumed	200	180	123	272	
Value Added	50	195	27		
Total Produced	250	375	150		

When this data is converted to a 3-company Leontief model, we get

$$A = \begin{pmatrix} 60/250 & 75/375 & 45/150 \\ 100/250 & 60/375 & 63/150 \\ 40/250 & 45/375 & 15/150 \end{pmatrix}, \quad D = \begin{pmatrix} 70 \\ 152 \\ 50 \end{pmatrix}.$$

a) Use A and D to calculate the production vector X for this model. How can you use the table to find X directly?

b) Find the new production vector if the demand for manufactured goods increases by 12% to 170. (Keep the technological constants the same.)

c) Repeat part b) if the demand for transportation increases by 12% to 60.

6. [88 & 89] Consider a model of the U.S. economy in 1958 as given by the table on the following page. (Source from which the data was derived: <u>Survey of Current Business</u> September, 1965.) Here we have aggregated all industries into 7 sectors, given in rows/columns numbered 1 through 7. The open sector is given in column 8. The value-added entry in each column represents the value which that sector adds to the products it consumes. The input-output matrix A and the demand matrix D are numbers 88 and 89.

a) Use A and D to calculate the production vector X for this model. How can you use the table to find X directly?

b) Find the new production vector if the demand for construction is increased by 10%. (Assume the technological constants are unchanged.)

c) Repeat part c) if the demand for services and trade is increased by 10%.

Input-Output Table for the U.S. Economy: 1958
(in Millions of Dollars)

↓ From/To →	1. Ag & For.	2. Min.	3. Const.	4. Mfg.	5. Trans. & Comm.
1. Agriculture & Forestry	15,578	0	237	25,065	37
2. Mining	103	1,129	756	13,346	29
3. Construction	613	10	8	762	1,550
4. Manufacturing	6,108	1,456	26,540	140,136	3,611
5. Transportation & Communication	983	534	2,231	12,944	2,545
6. Services & Trade	6,148	2,847	10,213	28,984	5,233
7. Other	1,091	1,956	379	13,985	2,555
Total	30,624	7,932	40,364	235,222	15,560
Value Added	22,098	10,416	28,937	126,304	31,352
Grand Total	52,722	18,348	69,301	361,526	46,912

↓ From/To →	6. Serv. & Trd.	7. Other	8. Final Demand	Total
1. Agriculture & Forestry	2,350	733	8,722	52,722
2. Mining	1,920	146	919	18,348
3. Construction	8,316	1,206	56,836	69,301
4. Manufacturing	24,462	5,407	153,806	361,526
5. Transportation & Communication	6,012	3,497	18,166	46,912
6. Services & Trade	48,806	2,341	177,500	282,072
7. Other	10,417	1,263	31,382	63,028
Total	102,283	14,593		893,909
Value Added	179,789	48,435		447,331
Grand Total	282,072	63,028	447,331	

Electrical Circuits

Simple electrical circuits can be analyzed by deriving a system of linear equations from the following two laws of circuits:

i) *Ohm's law:* The potential difference across a resistor is the product of the current and the value of the resistance.

ii) *Kirchhoff's current law:* The sum of the currents flowing out of a junction equals the sum of the currents flowing into the junction.

For example, in the circuit diagram below, we can find the unknown currents i_1, i_2, i_3 by noting that

$$i_1 = i_2 + i_3 \qquad \text{(Kirchhoff's current law)}$$

$$\text{and } v_1 - v_2 = 6i_3 \qquad \text{(Ohm's law)}$$

$$= 4i_2$$

$$= 12 - 2i_1.$$

Thus we have the system of equations

$$
\begin{aligned}
i_1 \quad - \quad i_2 \quad - \quad i_3 &= 0 \\
4i_2 \quad - \quad 6i_3 &= 0 \\
2i_1 \quad + \quad 4i_2 \qquad\quad &= 12 \,,
\end{aligned}
$$

which has the unique solution $i_1 = \frac{30}{11}$, $i_2 = \frac{18}{11}$, $i_3 = \frac{12}{11}$.

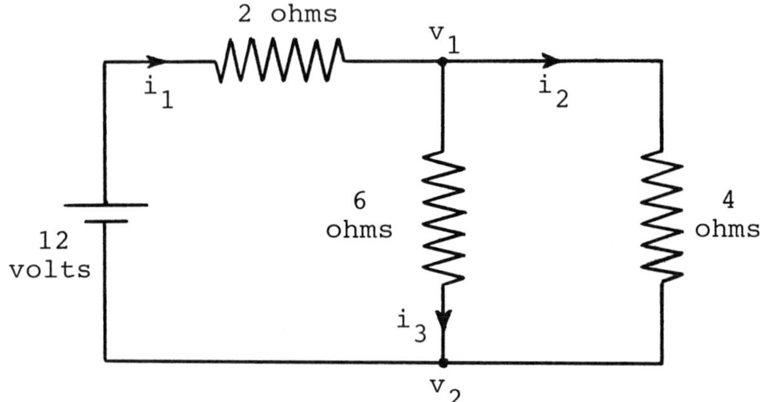

7. [70 & 71] Consider the given circuit diagram.

   a) Derive a system of equations that determines the currents. Write in matrix form as $AX = K$.
   Compare your results with matrices 70 and 71 in group storage.

   b) Find all the unknown currents.

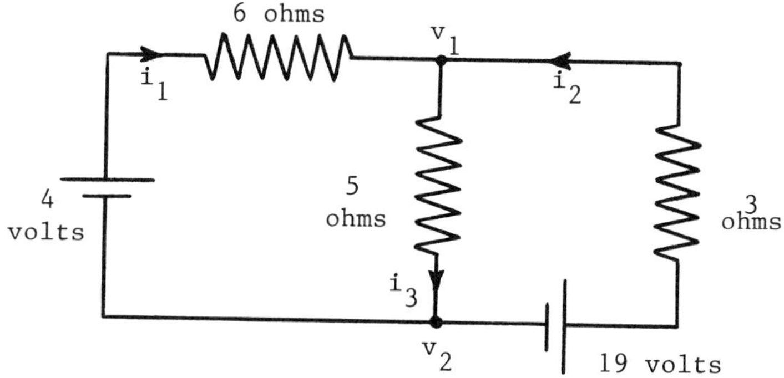

8. [90 & 91] Consider the given circuit diagram.
   Note that one resistor has unspecified resistance c.

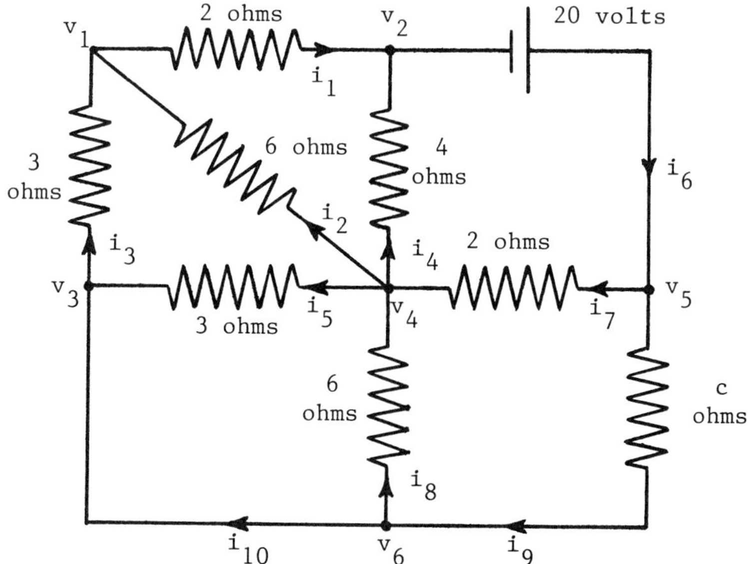

a) Derive a system of equations that determines the currents. Write in matrix form as $AX = K$.
Compare your results with matrices 90 and 91 in group storage. (Matrix 90 has $c = 2$.)

b) Find all the unknown currents if $c = 2$.
Find all the unknown currents if $c = 10$.
What do you notice about i_5 and i_8?
What do you notice about i_2?

c) There exists an integer value of c between 2 and 10 for which $i_5 = i_8 = 0$. Find this value of c.

Balancing equations for chemical reactions

Suppose we wish to balance the equation for a chemical reaction such as

$$C_2O_4^{2-} + MnO_2 + H^+ \rightarrow Mn^{2+} + CO_2 + H_2O.$$

An equation is *balanced* when:

i) the number of atoms of each chemical element involved is the same on each side of the equation, and

ii) the electrical charge is the same on each side of the equation.

We thus seek positive integers $x_1, x_2, x_3, x_4, x_5, x_6$, with no common divisors larger than 1, so that the general equation

$$(x_1)C_2O_4^{2-} + (x_2)MnO_2 + (x_3)H^+ \rightarrow (x_4)Mn^{2+} + (x_5)CO_2 + (x_6)H_2O$$

satisfies conditions i) and ii). This leads to a system of linear equations in the unknowns x_i:

$$
\begin{array}{rrcl}
\text{(C)} & & 2x_1 & = x_5 \\
\text{(O)} & & 4x_1 + 2x_2 & = 2x_5 + x_6 \\
\text{(Mn)} & & x_2 & = x_4 \\
\text{(H)} & & x_3 & = 2x_6 \\
\text{(electrical} & & -2x_1 + x_3 & = 2x_4 \\
\text{charge)} & & &
\end{array}
$$

$$
\text{or} \quad AX = 0 \quad \text{where} \quad A = \begin{pmatrix} 2 & 0 & 0 & 0 & -1 & 0 \\ 4 & 2 & 0 & 0 & -2 & -1 \\ 0 & 1 & 0 & -1 & 0 & 0 \\ 0 & 0 & 1 & 0 & 0 & -2 \\ -2 & 0 & 1 & -2 & 0 & 0 \end{pmatrix}.
$$

We use MAX to find a basis for the nullspace of A and thus solve the problem.

```
Command:  get 101 a
Get matrix 101 and store it in A
      2.         0.         0.         0.        -1.         0.
      4.         2.         0.         0.        -2.        -1.
      0.         1.         0.        -1.         0.         0.
      0.         0.         1.         0.         0.        -2.
     -2.         0.         1.        -2.         0.         0.
Command: rank a
Rank -- record the rank of A

The following upper-triangular matrix has the same rank as A:
     -4.9       -0.82       0.82      -1.63        2.04        0.41
                 2.08       0.32      -1.12        0.8        -0.8
                           -2.06       0.15 -3.74E-02          1.01
                                       1.03        0.23       -0.37
                                                  -0.37        0.19
Enter the rank of A (it is at most 5): 5
Command: nbasis a
NBasis -- find a nullspace basis of A and store in %
      0.25
      0.25
      1.
      0.25
      0.5
      0.5
Command: compute 4*%
Compute (4*%) and store in %
      1.
      1.
      4.
      1.
      2.
      2.
```

Hence the balanced reaction is

$$C_2O_4^{2-} + MnO_2 + 4H^+ \rightarrow Mn^{2+} + 2CO_2 + 2H_2O.$$

Balance the following chemical reactions.

9. [102] $C_2O_4^{2-} + MnO_4^- + H^+ \rightarrow Mn^{2+} + CO_2 + H_2O$

10. [103] $Cr_2O_7^{2-} + I^- + H^+ \rightarrow Cr^{3+} + I_2 + H_2O$

11. [104] $ClO_3^- + C_{12}H_{22}O_{11} \rightarrow CO_2 + H_2O + Cl^-$

12. [105] $CuS + NO_3^- + H^+ \rightarrow Cu^{2+} + S + NO + H_2O$

13. [106] $CH_2O + Ag(NH_3)_2^+ + OH^- \rightarrow HCO_3^- + Ag + NH_3 + H_2O$

14. [107] $H_2O_2 + MnO_4^- + H^+ \rightarrow Mn^{2+} + H_2O + O_2$

15. [108] $C_2H_5OH + H_2O + I_3^- \rightarrow CO_2 + CHO_2^- + HCI_3 + I^- + H^+$

Flow networks

A system of water pipes is connected at points A, B, C, D, E, F as shown in the diagram:

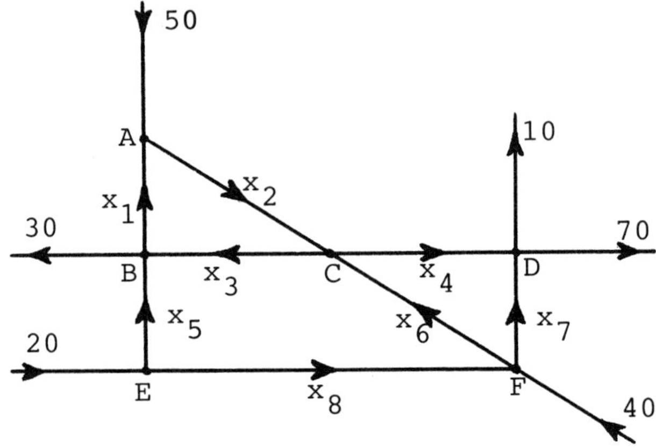

The numbers represent flow rates (in gal/min) in and out of the system with arrows indicating the direction of the flow. The flow at the connections can be controlled by valves.

Suppose it is necessary to repair the pipe between connections C and D. In this case we wish to have as small a flow as possible through this section of pipe. What is this minimum flow? What are the flows through the other sections of pipe when this minimum is attained?

Label the unknown flows x_1 through x_8 as in the diagram. (Since no flow can be reversed, each x_i is nonnegative.) A steady flow requires that the total amount of water entering a connection must equal the amount leaving it. This conservation of flow constraint leads to 6 equations in 8 unknowns. For example, the equation at connection A is $x_1 + 50 = x_2$.

16. [109 & 110] a) Derive the equations at connections B through F. Write the equations in matrix form and compare your answer with matrices 109 and 110 in group storage.

   b) Use MAX to find the general solution of this system of equations.

   c) Find the minimal nonnegative value of x_4. For this minimal value of x_4, find all possible values for the remaining x_i.

   d) Now suppose it is necessary to repair the pipe between connections B and C. Find the minimal nonnegative value of x_3. For this minimum value, find all possible values for the remaining x_i.

   e) Repeat parts a) through c) if all flows in and out of this system are increased by 50%.

Graphs

A *graph* G consists of a finite collection of *vertices* (points) and *edges* (lines or arcs) joining different pairs of distinct vertices. The figure shows a graph with 4 vertices and 5 edges.

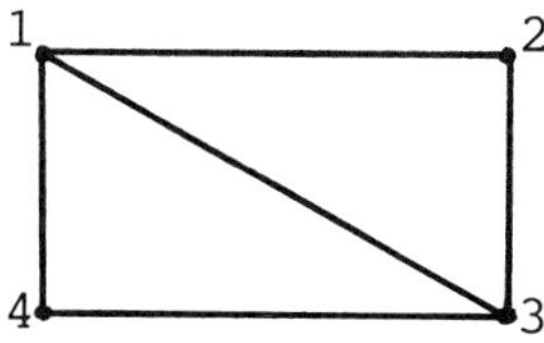

Let G be a graph with n vertices labeled 1 through n. The *adjacency matrix* of G is the n by n matrix $A = \left(a_{ij} \right)$ where

$$a_{ij} = \begin{cases} 1, & \text{if an edge joins vertices } i \text{ and } j \\ 0, & \text{otherwise.} \end{cases}$$

The adjacency matrix of the above graph is $A = \begin{pmatrix} 0 & 1 & 1 & 1 \\ 1 & 0 & 1 & 0 \\ 1 & 1 & 0 & 1 \\ 1 & 0 & 1 & 0 \end{pmatrix}$.

In a *directed graph*, the edges are directed by arrows, as in the figure below.

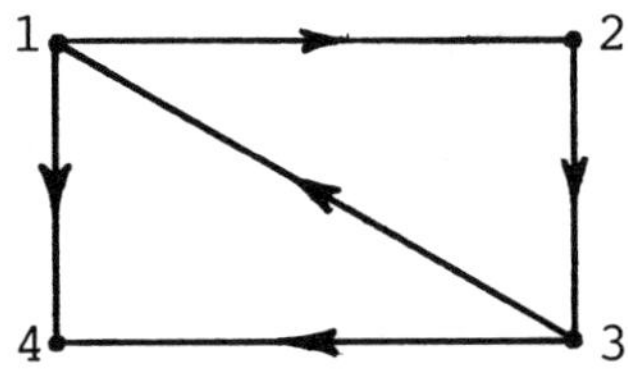

The adjacency matrix of a directed graph with vertices 1 through n is $A = \left(a_{ij} \right)$ where

$$a_{ij} = \begin{cases} 1, & \text{if there is a directed edge from } i \text{ to } j \\ 0, & \text{otherwise.} \end{cases}$$

For the above directed graph, $A = \begin{pmatrix} 0 & 1 & 0 & 1 \\ 0 & 0 & 1 & 0 \\ 1 & 0 & 0 & 1 \\ 0 & 0 & 0 & 0 \end{pmatrix}$. The transpose

$A' = \begin{pmatrix} 0 & 0 & 1 & 0 \\ 1 & 0 & 0 & 0 \\ 0 & 1 & 0 & 0 \\ 1 & 0 & 1 & 0 \end{pmatrix}$ is the adjacency matrix of the directed graph with all edges reversed as shown below.

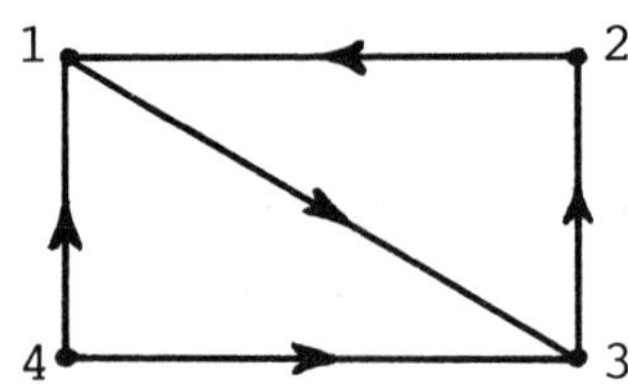

Let A be the adjacency matrix of a graph. The (i,j) entry of its square A^2 gives the number of different 2-edge paths between vertices i and j. Similarly for higher powers of A. Thus the (i,j) entry in $A + A^2 + A^3$ is the number of paths between vertices i and j that have at most 3 edges.

17. [111] Consider the following road map of central Iowa. A traveling salesperson wishes to schedule round-trip routes that include at most three intermediate stops. (It is possible for a route to stop in the same city more than once.)

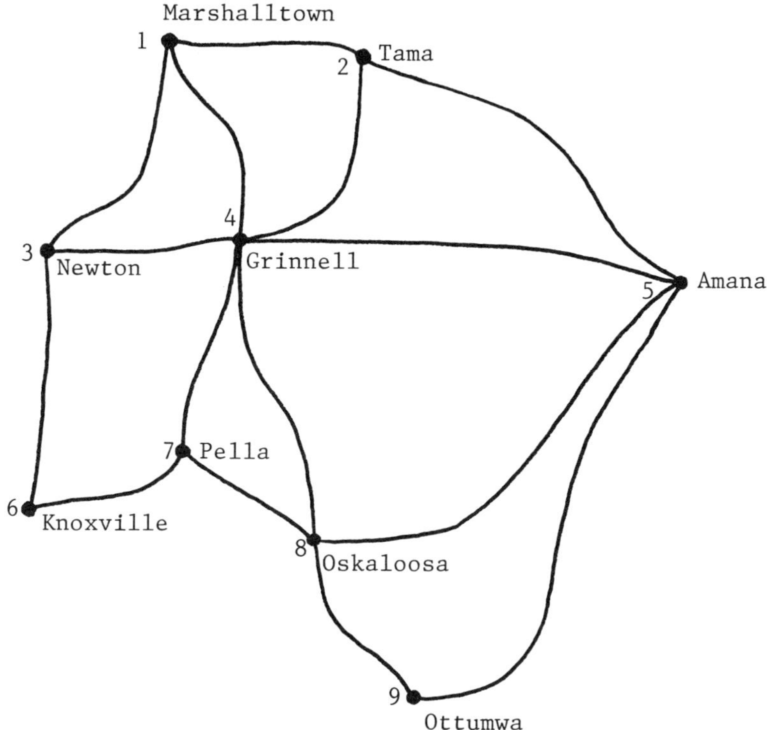

a) Find the adjacency matrix for this graph.
   Compare your answer with matrix 111.

b) Find the number of such routes possible for a salesperson based at Grinnell; at Amana; at Ottumwa.

18. [112] Consider the well-known missionaries/cannibals puzzle: Three missionaries and three cannibals wish to cross a river (say from the left shore to the right shore) using a two-person boat. At no time can the cannibals outnumber the missionaries on either side of the river (unless no missionaries are present on a shore). Devise a series of crossings to accomplish this task.

Suppose we revise the problem as follows: Find the minimal number of crossings (one-way trips across the river). Find the number of ways this minimal number of crossings can be achieved.

We can use a directed graph to help solve the revised problem. Denote a position of the six people by an ordered pair (a, b) where a (resp. b) is the number of missionaries (resp. cannibals) on the left shore. There are 10 allowable positions: (3,3) (the starting position), (3,2), (3,1), (3,0), (2,2), (1,1), (0,3), (0,2), (0,1), (0,0) (the end position). Let these pairs be vertices 1 through 10 of a directed graph; draw a directed edge from vertex i to vertex j if a crossing exists from the left shore to the right shore that changes position i to position j.

a) Find the adjacency matrix A for this directed graph.
   Compare your answer with matrix 112.

b) Use A and its transpose A' to find the minimal number of crossings and the number of ways this minimal number of crossings can be achieved.

Chapter 2

Vectors and Vector Spaces

Section 2.1   SPAN

Given vectors $v_1, \ldots, v_n$ and w in a vector space **V**, we wish to determine whether or not w lies in the subspace of **V** spanned by $v_1, \ldots, v_n$. To do this, we need to decide if constants $c_1, \ldots, c_n$ exist so that $w = c_1 v_1 + \ldots + c_n v_n$. This often leads to deciding whether a system $AX = K$ has a solution. From Section 1.4, we know how to do this:

$$AX = K \text{ has a solution iff rank } A = \text{rank} \begin{pmatrix} A & K \end{pmatrix}.$$

Example: Determine whether or not the polynomial $x^2 + x + 1$ lies in the subspace of polynomials spanned by $x^3 - 2x, x^3 + 2x^2 - 3, x^3 + x^2 - x - 2$.

We look for constants c_1, c_2, c_3 so that

$$x^2 + x + 1 = c_1(x^3 - 2x) + c_2(x^3 + 2x^2 - 3) + c_3(x^3 + x^2 - x - 2).$$

Equating coefficients gives

$$\begin{aligned}
0 &= c_1 + c_2 + c_3 \\
1 &= 2c_2 + c_3 \\
1 &= -2c_1 - c_3 \\
1 &= -3c_2 - 2c_3
\end{aligned}$$

or
$$\begin{pmatrix} 1 & 1 & 1 \\ 0 & 2 & 1 \\ -2 & 0 & -1 \\ 0 & -3 & -2 \end{pmatrix} \begin{pmatrix} c_1 \\ c_2 \\ c_3 \end{pmatrix} = \begin{pmatrix} 0 \\ 1 \\ 1 \\ 1 \end{pmatrix}.$$

```
Command:  get 72 a
Get matrix 72 and store it in A
      1.          1.          1.
      0.          2.          1.
     -2.          0.         -1.
      0.         -3.         -2.
Command:  get 73 k
Get matrix 73 and store it in K
      0.
      1.
      1.
      1.
Command:  aug a k ak
Augment A by K and store the result in AK
      1.          1.          1.          0.
      0.          2.          1.          1.
     -2.          0.         -1.          1.
      0.         -3.         -2.          1.
Command:  rank a
Rank -- record the rank of A

The following upper-triangular matrix has the same rank as A:
    -3.74       -0.27       -2.41
                 2.22        1.06
                            -0.29

Enter the rank of A (it is at most 3): 3
Command:  rank ak
Rank -- record the rank of AK

The following upper-triangular matrix has the same rank as AK:
    -3.74       -0.27        0.27       -2.41
                 2.22       -0.87        1.06
                             1.47       -0.29
                                       -5.96E-08

Determinant =  0.00E0     r-condition =        0.

Enter the rank of AK (it is at most 4):  3
```

Hence $x^2 + x + 1$ does lie in the given subspace.

Command: **solve a k**
Solve the system A K and store the solution in %

    2.

    3.

    -5.

WARNING: This is a least-squares solution, which need not be an
         actual solution

This computation shows

$$x^2 + x + 1 = 2(x^3 - 2x) + 3(x^3 + 2x^2 - 3) - 5(x^3 + x^2 - x - 2),$$

as is easily checked.

Suppose we ask if the polynomial $x^2 + 4x + 1$ lies in the given subspace.
Keep A and form a new augmented matrix AK. We know rank $A = 3$ and
find rank $AK = 4$. Hence $x^2 + 4x + 1$ does not lie in the subspace spanned
by $x^3 - 2x, x^3 + 2x^2 - 3$ and $x^3 + x^2 - x - 2$.

PROBLEMS

1. [74, 75] Determine whether or not each of the following vectors in $\mathbf{R}^4$
   belongs to the subspace spanned by $(-1, 3, 0, 2)$, $(5, 1, 1, -3)$, $(3, 7, 1, 1)$,
   $(0, 1, -2, 4)$.

   a) $(7, 12, 0, 4)$

   b) $(1, 1, 0, 1)$

   c) $(-17, 7, -11, 29)$

2. [76, 77] Determine whether or not each of the following polynomials
   belongs to the subspace spanned by $x^3 + 2x^2 + 1, x^3 + x, x^2 - 2$.

   a) $x^2 - x - 3$

   b) $x^2 - x + 3$

   c) $4x^3 - 3x - 5$

   d) $x - 5$

3. [78] Show that the following two collections of vectors span the same subspace of $\mathbf{R}^4$.

   a) $(1, -1, 1, 6), (-3, 2, 1, -5), (0, -3, 12, 1)$

   b) $(1, -3, 9, -6), (9, -5, -7, 2), (-11, 9, -3, -2)$

4. [79] Determine whether or not $\begin{pmatrix} 3 & 1 \\ -6 & -4 \end{pmatrix}$ lies in the subspace of the space of 2 by 2 matrices spanned by

$$\begin{pmatrix} 1 & 7 \\ 2 & 1 \end{pmatrix}, \begin{pmatrix} -4 & 0 \\ 12 & -1 \end{pmatrix}, \begin{pmatrix} 3 & 6 \\ -5 & 2 \end{pmatrix}.$$

Section 2.2 LINEAR INDEPENDENCE

Suppose we wish to test linear independence of vectors $v_1, \ldots, v_n$ in a vector space $\mathbf{V}$. Using the definition, we assume there exist constants $c_1, \ldots, c_n$ such that $c_1 v_1 + \ldots + c_n v_n = 0$ and determine whether or not this forces $c_1 = \ldots = c_n = 0$. The vector equation $c_1 v_1 + \ldots + c_n v_n = 0$ often leads to a matrix equation of the form $AC = 0$ where $C = \begin{pmatrix} c_i \end{pmatrix}$. In this case, linear independence of $v_1, \ldots, v_n$ depends on the matrix A:

$$v_1, \ldots, v_n \text{ are linearly independent}$$
$$\text{iff the nullspace of } A \text{ consists of the zero vector only}$$
$$\text{iff } A \text{ has rank } n.$$

Example 1: Determine whether the vectors $(1, -1, 1)$, $(-3, 2, 1)$, $(2, 3, -4)$ in $\mathbf{R}^3$ are linearly independent.

Suppose $c_1(1, -1, 1) + c_2(-3, 2, 1) + c_3(2, 3, -4) = (0, 0, 0)$. Equating coordinates yields

$$
\begin{array}{rrrrl}
c_1 & - \ 3c_2 & + \ 2c_3 & = & 0 \\
-c_1 & + \ 2c_2 & + \ 3c_3 & = & 0 \\
c_1 & + \ c_2 & - \ 4c_3 & = & 0
\end{array}
$$

$$
\text{or} \quad
\begin{pmatrix} 1 & -3 & 2 \\ -1 & 2 & 3 \\ 1 & 1 & -4 \end{pmatrix}
\begin{pmatrix} c_1 \\ c_2 \\ c_3 \end{pmatrix}
=
\begin{pmatrix} 0 \\ 0 \\ 0 \end{pmatrix}.
$$

```
Command:  get 80 a
Get matrix 80 and store it in A
      1.          -3.           2.
     -1.           2.           3.
      1.           1.          -4.
Command:  rank a
Rank -- record the rank of A

The following upper-triangular matrix has the same rank as A:
    -5.39         0.74          0.93
                 -3.67          1.28
                               0.71
Determinant = -1.40E1      r-condition =  8.52E-02
Enter the rank of A (it is at most 3):  3
```

Hence we conclude that the three given vectors are linearly independent. We confirm this using the **nbasis** command.

```
Command: nbasis a
NBasis -- find a nullspace basis of A and store in %
The nullspace is zero. No result is stored.
```

Since the matrix equation has only the solution $c_1 = c_2 = c_3 = 0$, again we conclude that the vectors are linearly independent.

Example 2: Determine whether or not the vectors $1, 3x, e^{-x}, e^{2x}$ in the vector space of real-valued functions are linearly independent.

Suppose $c_1 1 + c_2(3x) + c_3 e^{-x} + c_4 e^{2x} = 0$.

Differentiate both sides: $3c_2 - c_3 e^{-x} + 2c_4 e^{2x} = 0$.

Differentiate again: $c_3 e^{-x} + 4c_4 e^{2x} = 0$.

And again: $-c_3 e^{-x} + 8c_4 e^{2x} = 0$.

Set $x = 0$ in each of the four equations:

$$\begin{aligned}
c_1 + c_3 + c_4 &= 0 \\
3c_2 - c_3 + 2c_4 &= 0 \\
c_3 + 4c_4 &= 0 \\
-c_3 + 8c_4 &= 0
\end{aligned}$$

$$\text{or } AC = 0 \quad \text{where} \quad A = \begin{pmatrix} 1 & 0 & 1 & 1 \\ 0 & 3 & -1 & 2 \\ 0 & 0 & 1 & 4 \\ 0 & 0 & -1 & 8 \end{pmatrix}.$$

We check that the rank of A is 4 (equivalently, the nullspace of A contains only the zero vector) and hence the functions are linearly independent.

PROBLEMS

1. Determine whether or not the following collections of vectors in $\mathbf{R}^4$ are linearly independent.

   a) [6] $(1, -1, 2, 0), (1, 1, 2, 0), (3, 0, 0, 1)$

   b) [7] $(1, 2, -1, 2), (-2, -5, 3, 0), (1, 0, 1, 10)$

  c) [74] $(-1,3,0,2),(5,1,1,-3),(3,7,1,1),(0,1,-2,4)$

  d) [35] $(1,1,0,1),(1,0,1,0),(5,-3,2,7),(0,6,0,-5)$

  e) [128] $(1,1,0,1),(1,0,1,0),(5,-3,2,7),(0,6,0,-5),$
      $(-2,5,-1,-5)$

2. Determine whether or not the following collections of polynomials are linearly independent.

  a) [121] $x^2 - 1, x^2 + x - 2, x^2 + 3x + 2$

  b) [151] $x^2 - 1, x^2 + x - 2, x^2 + 3x + 2, x^2 + 4$

  c) [122] $x^3 + 2x + 5, x^3 - 3x^2 + 3, 3x^2 + 2, 6x$

3. Determine whether or not the following collections of functions are linearly independent.

  a) $1, \sin x, \cos x$

  b) $1, \sin x, \sin 2x$

  c) $\sin^2 x, \cos^2 x, \cos 2x$

  d) e^x, e^{-2x}, e^{3x}

  e) $e^x, xe^x, x^2 e^x$

4. Think of the following vectors in $\mathbf{R}^3$ as directed line segments with their initial points at the origin. Determine whether or not each collection lies in a plane.

  a) [123] $(3,0,1),(-1,-2,3),(7,2,-1)$

  b) [66] $(3,3,-4),(-2,-1,-3),(0,5,-6)$

5. [7] Determine whether or not the following collection of 2 by 2 matrices is linearly independent.

$$\begin{pmatrix} 1 & 2 \\ -1 & 2 \end{pmatrix}, \begin{pmatrix} -2 & -5 \\ 3 & 0 \end{pmatrix}, \begin{pmatrix} 1 & 0 \\ 1 & 10 \end{pmatrix}$$

Section 2.3 BASIS

Vectors $v_1, \ldots, v_n$ in a vector space $\mathbf{V}$ form a *basis* for $\mathbf{V}$ if they both span $\mathbf{V}$ and are linearly independent. The *dimension* of $\mathbf{V}$ is the number of vectors in a basis. Note that if $\mathbf{V}$ has dimension n, then n vectors in $\mathbf{V}$ form a basis if they either span $\mathbf{V}$ or are linearly independent.

Example 1: The vectors $(1, 2, -3)$, $(3, -4, 5)$, $(13, -4, 3)$, $(-5, 5, 6)$ span $\mathbf{R}^3$. Find a basis for $\mathbf{R}^3$ from among these four vectors.

Arrange the vectors as columns of a 3 by 4 matrix A and find a basis for the column space of A, i.e., the space spanned by the columns of A.

```
Command: get 124 a
Get matrix 124 and store it in A
        1.        3.        13.        -5.
        2.       -4.        -4.         5.
       -3.        5.         3.         6.
Command: cbasis a c

Enter the rank of A (it is at most 3): 3
CBasis -- find a columnspace basis of A and store in C
       13.       -5.        3.
       -4.        5.       -4.
        3.        6.        5.
```

Hence the vectors $(13, -4, 3), (-5, 5, 6), (3, -4, 5)$ form a basis for $\mathbf{R}^3$. To express the omitted vector $(1, 2, -3)$ as a linear combination of the basis vectors, use the **solve** command.

```
Command:  solve c a(,1)

Enter the rank of C (it is at most 3): 3
Solve the system C A(1-3,1-1) and store the solution in %
       0.25
       0.
      -0.75
```

Hence $(1, 2, -3) = \frac{1}{4}(13, -4, 3) - \frac{3}{4}(3, -4, 5)$.

Example 2: Let A be the 4 by 5 matrix 125 in group storage. Find bases for the nullspace, column space, and row space of A.

We remind ourselves: The nullspace of A is the subspace of $\mathbf{R}^5$ of vectors X such that $AX = 0$; the column space is the subspace of $\mathbf{R}^4$ spanned by the columns of A; the row space is the subspace of $\mathbf{R}^5$ spanned by the rows of A.

```
Command:  get 125 a
Get matrix 125 and store it in A
     4.        0.        1.        1.        0.
     3.        0.        0.        1.       -1.
     0.        1.        2.        0.        1.
     5.       -2.        1.        0.        3.
Command: rank a
Rank -- record the rank of A
The following upper-triangular matrix has the same rank as A:
   -7.07      -1.7      -1.27      1.41      -0.99
               2.85      1.       -0.91      -0.94
                        -1.84     -1.47      -0.37
                                   5.96E-08   0.
Enter the rank of A (it is at most 4):  3
Command:  nbasis a
NBasis -- find a nullspace basis of A and store in %
     0.2       -0.2
     1.         0.                    .
    -0.8       -0.2
     0.         1.
     0.6        0.4
Command:  cbasis a
CBasis -- find a columnspace basis of A and store in %
     4.         0.        1.
     3.        -1.        0.
     0.         1.        2.
     5.         3.        1.
```

To find a basis for the row space of A, we find a basis for the column space of the transpose A' of A.

```
Command:  transpose a ap#
```
Transpose A and store the transpose in AP

```
Command:  rank ap
```
Rank -- record the rank of AP

The following upper-triangular matrix has the same rank as AP:

$$
\begin{array}{cccc}
-6.24 & -1.92 & -0.48 & -3.36 \\
 & -2.7 & 0.71 & -2.42 \\
 & & -2.29 & -0.92 \\
 & & & 0.
\end{array}
$$

Enter the rank of AP (it is at most 4): **3**

```
Command:  cbasis ap
```
CBasis -- find a columnspace basis of AP and store in %

$$
\begin{array}{ccc}
5. & 3. & 0. \\
-2. & 0. & 1. \\
1. & 0. & 2. \\
0. & 1. & 0. \\
3. & -1. & 1.
\end{array}
$$

PROBLEMS

1. For each of the following collections of vectors $\{v_1, \ldots, v_m\}$, find the dimension of the subspace S spanned by the vectors, select a basis for S from the set $\{v_1, \ldots, v_m\}$, and express each of the remaining v_i as a linear combination of the basis vectors.

   a) [54] $(-1, 3, 7), (5, -2, 1), (7, 5, 23), (1, 10, 29), (2, 7, 22)$

   b) [56] $(-1, 1, 2, 5), (3, -2, 0, 1), (-5, 4, 4, 9), (4, 0, -3, 2),$
   $(18, -7, -8, 2)$

2. Repeat Problem 1 for each of the following collections of polynomials.

   a) [126] $2x^3 + 5x^2 - x + 3, x^3 - 6x - 5, 7x^3 + 9x^2 - 3x + 4, 17x^3 - 6x - 7$

   b) [127] $x^3 - 3x^2 + x + 7, x^3 - 2x + 4, x^3 + 3x^2 - 5x + 1,$
   $2x^3 + x^2 - 5x + 11, 4x^3 - 5x^2 - 3x + 25$

3. [50] For the indicated matrix A, do the following parts.

   a) Find a collection of columns of A which form a basis of the column space of A.

   b) Express each column of A as a linear combination of the basis vectors in a).

   c) Repeat parts a) and b) for the rows of A instead of the columns of A.

4. [50, 49, 128] For each of the indicated matrices A, do the following parts.

   a) Find the six numbers:

      rank of A

      dimension of the column space of A

      dimension of the nullspace of A

      dimension of the row space of A

      rank of A'

      dimension of the nullspace of A'

   b) State all the relationships that you believe hold among the six numbers in part a) for any matrix A.

   c) Prove your assertions in part b).

5. [67] Determine whether or not each of the following collections of matrices forms a basis for the space of all 2 by 2 matrices.

   a) $\begin{pmatrix} 1 & 2 \\ 3 & -4 \end{pmatrix}, \begin{pmatrix} 3 & 0 \\ -2 & 2 \end{pmatrix}, \begin{pmatrix} 7 & -1 \\ -1 & 1 \end{pmatrix}, \begin{pmatrix} 10 & 5 \\ -3 & 0 \end{pmatrix}$

   b) $\begin{pmatrix} 1 & 2 \\ 3 & -4 \end{pmatrix}, \begin{pmatrix} 3 & 0 \\ -2 & 2 \end{pmatrix}, \begin{pmatrix} 7 & -1 \\ -1 & 1 \end{pmatrix}, \begin{pmatrix} 10 & 5 \\ -3 & 1 \end{pmatrix}$

6. Let $v_1, \ldots, v_n$ be a basis for a vector space $\mathbf{V}$.

   a) For what values of n do $v_1 + v_2, v_2 + v_3, \ldots, v_{n-1} + v_n, v_n + v_1$ form a basis for $\mathbf{V}$? Answer this by trying small values of n.

   b) Prove that your answer in part a) is correct for all n.

Section 2.4 CHANGE OF BASIS

Given a basis $\mathcal{B} = \{v_1, \ldots, v_n\}$ for a vector space $\mathbf{V}$, a vector w in $\mathbf{V}$ can be written uniquely as $w = a_1 v_1 + \ldots + a_n v_n$. The scalars $a_1, \ldots, a_n$ are the *coordinates of w with respect to the basis $\mathcal{B}$*; we write these coordinates as a column vector:

$$[w]_{\mathcal{B}} = \begin{pmatrix} a_1 \\ \vdots \\ a_n \end{pmatrix}.$$

Given two bases $\mathcal{B} = \{v_1, \ldots, v_n\}$ and $\mathcal{B}' = \{v_1', \ldots, v_n'\}$ for $\mathbf{V}$, a vector w in $\mathbf{V}$ has coordinates with respect to each basis: $[w]_{\mathcal{B}}$ and $[w]_{\mathcal{B}'}$. What is the relationship between these column vectors? Answer: There exists an invertible n by n matrix P such that

$$[w]_{\mathcal{B}} = P[w]_{\mathcal{B}'}.$$

Indeed, P is the matrix whose columns are the coordinates of the v_i' with respect to $\mathcal{B}$, i.e.,

$$P = \left([v_1']_{\mathcal{B}} \cdots [v_n']_{\mathcal{B}} \right).$$

(Proof: If $w = a_1 v_1 + \ldots + a_n v_n = a_1' v_1' + \ldots + a_n' v_n'$, then

$$[w]_{\mathcal{B}} = [a_1' v_1' + \ldots + a_n' v_n']_{\mathcal{B}} = a_1' [v_1']_{\mathcal{B}} + \ldots + a_n' [v_n']_{\mathcal{B}}$$

$$= \left([v_1']_{\mathcal{B}} \cdots [v_n']_{\mathcal{B}} \right) \begin{pmatrix} a_1' \\ \vdots \\ a_n' \end{pmatrix} = P[w]_{\mathcal{B}'}.)$$

P is the *transition matrix from $\mathcal{B}'$ to $\mathcal{B}$*. From $[w]_{\mathcal{B}} = P[w]_{\mathcal{B}'}$ we get $[w]_{\mathcal{B}'} = P^{-1}[w]_{\mathcal{B}}$ and P^{-1} is the transition matrix from $\mathcal{B}'$ to $\mathcal{B}$, i.e.,

$$P^{-1} = \left([v_1]_{\mathcal{B}'} \cdots [v_n]_{\mathcal{B}'} \right).$$

Example: Let $\mathcal{B} = \{(1,0,0), (0,1,0), (0,0,1)\}$ and $\mathcal{B}' = \{(-1,3,7), (4,-2,0), (2,1,6)\}$, which are bases for $\mathbf{R}^3$.

a) Find the coordinates of $(-2, 5, 10)$ with respect to $\mathcal{B}'$.

b) Find the transition matrices from $\mathcal{B}'$ to $\mathcal{B}$ and from $\mathcal{B}$ to $\mathcal{B}'$.

For a), we seek a_1, a_2, a_3 so that

$$(-2, 5, 10) = a_1(-1, 3, 7) + a_2(4, -2, 0) + a_3(2, 1, 6), \quad \text{i.e.,}$$

$$\begin{aligned} -2 &= -a_1 + 4a_2 + 2a_3 \\ 5 &= 3a_1 - 2a_2 + a_3 \\ 10 &= 7a_1 \qquad\quad + 6a_3. \end{aligned}$$

Using the **solve** command, we find the solution of the matrix equation

$$\begin{pmatrix} -1 & 4 & 2 \\ 3 & -2 & 1 \\ 7 & 0 & 6 \end{pmatrix} \begin{pmatrix} a_1 \\ a_2 \\ a_3 \end{pmatrix} = \begin{pmatrix} -2 \\ 5 \\ 10 \end{pmatrix}$$

to be $\begin{pmatrix} 4 \\ 2 \\ -3 \end{pmatrix}$. Hence $[(-2, 5, 10)]_{\mathcal{B}'} = \begin{pmatrix} 4 \\ 2 \\ -3 \end{pmatrix}$.

In b), the transition matrix from $\mathcal{B}'$ to $\mathcal{B}$ is easy: The coordinates of any vector with respect to the standard basis $\mathcal{B}$ are given directly. Thus

$$P = \left([v_1']_{\mathcal{B}} \; [v_2']_{\mathcal{B}} \; [v_3']_{\mathcal{B}} \right) = \begin{pmatrix} -1 & 4 & 2 \\ 3 & -2 & 1 \\ 7 & 0 & 6 \end{pmatrix}.$$

The transition matrix from $\mathcal{B}$ to $\mathcal{B}'$ is

$$P^{-1} = \begin{pmatrix} 3 & 6 & -2 \\ 11/4 & 5 & -7/4 \\ -7/2 & -7 & 5/2 \end{pmatrix}.$$

(Use the **invert** command.)

From this result, we can check our answer in a):

$$\begin{pmatrix} 3 & 6 & -2 \\ 11/4 & 5 & -7/4 \\ -7/2 & -7 & 5/2 \end{pmatrix} \begin{pmatrix} -2 \\ 5 \\ 10 \end{pmatrix} = \begin{pmatrix} 4 \\ 2 \\ -3 \end{pmatrix}.$$

(Use the **multiply** or **compute** command.)

PROBLEMS

1. [129, 130] Let $B = \{(1, 4, -1), (-3, 2, 1), (-3, 0, 3)\}$ and $B' = \{(3, 5, 4), (-2, 0, 1), (1, -1, 2)\}$, which are bases for $\mathbf{R}^3$.

   a) Find the transition matrix P from B' to B.
   Find the transition matrix Q from B to B'.

   b) For $w = (-21, -1, 13)$ in $\mathbf{R}^3$, find $[w]_B$ and $[w]_{B'}$. Verify that $[w]_B = P[w]_{B'}$ and $[w]_{B'} = Q[w]_B$.

2. [131, 132] Repeat Problem 1 for the bases
   $B = \{(1, 1, 1, -1), (1, 1, -1, 1), (1, -1, 1, 1), (-1, 1, 1, 1)\}$ and
   $B' = \{(1, 2, -1, 0), (3, 1, -1, 2), (0, 3, 2, 1), (-5, 4, -7, 2)\}$ for $\mathbf{R}^4$.
   In part b), let $w = (3, 5, 19, -1)$.

3. [133, 134] Repeat Problem 1 for $B = \{x^3 + 1, x^2 + 1, x + 1, 1\}$ and $B' = \{x^3 + x, x^3 - x, x^2 + 1, x^2 - 1\}$, which are bases for the space of polynomials of degree less than 4.
   In part b), let $w = x^3 + x^2 + 2x + 1$.

4. [135] Repeat Problem 1 for the bases $B = \{x^4, x^3, x^2, x, 1\}$ and $B' = \{(x + 1)^4, (x + 1)^3, (x + 1)^2, x + 1, 1\}$ of the space of polynomials of degree less than 5.
   In part b), let $w = x^4 - x^3 + x^2 - x - 1$.

5. [136] Find the coordinates of $\begin{pmatrix} 2 & 2 \\ -1 & 4 \end{pmatrix}$ with respect to the basis
   $$B = \left\{ \begin{pmatrix} 1 & -8 \\ 6 & 10 \end{pmatrix}, \begin{pmatrix} 2 & 0 \\ -1 & 5 \end{pmatrix}, \begin{pmatrix} -3 & -2 \\ 1 & 4 \end{pmatrix}, \begin{pmatrix} -4 & 1 \\ -1 & 0 \end{pmatrix} \right\}$$
   of the space of all 2 by 2 matrices.

Section 2.5 ORTHONORMAL BASES

Denote the standard inner (dot) product of vectors v and w in $\mathbf{R}^n$ by $v \cdot w$ and denote the length of v by $\| v \|$. Vectors $u_1, \ldots, u_r$ in $\mathbf{R}^n$ are *orthonormal* if they are orthogonal unit vectors, i.e., if

$$u_i \cdot u_j = \begin{cases} 1 & \text{if } i = j \\ 0 & \text{if } i \neq j \end{cases}.$$

Note that if we write vectors v and w as n by 1 matrices V and W, then $v \cdot w = V'W$.

Let $v_1, \ldots, v_r$ be linearly independent vectors in $\mathbf{R}^n$. We can produce a set of orthonormal vectors $u_1, \ldots, u_r$ that span the same subspace as $v_1, \ldots, v_r$ using the Gram-Schmidt process. An algorithm for producing the u's by hand is as follows:

Let $\quad u_1 = \frac{1}{\|v_1\|} v_1$.

Assuming $\quad u_1, \ldots, u_{i-1}$ have been determined,

let $\quad w_i = v_i - (v_i \cdot u_1)u_1 - (v_i \cdot u_2)u_2 - \ldots - (v_i \cdot u_{i-1})u_{i-1}$

and let $\quad u_i = \frac{1}{\|w_i\|} w_i$.

Let $A = \begin{pmatrix} v_1 \ldots v_r \end{pmatrix}$ be an n by r matrix with linearly independent columns. Gram-Schmidt produces an orthonormal basis $u_1, \ldots, u_r$ for the column space of A. Considered as a whole, the process yields a factorization of A (just as Gaussian elimination yielded an LU factorization of a nonsingular matrix). The algorithm described above gives a system of equations that can be solved for the v's in terms of the u's; in fact, we get v_i in terms of $u_1, \ldots, u_i$. Hence

$$\begin{pmatrix} v_1 \ldots v_r \end{pmatrix} = \begin{pmatrix} u_1 \ldots u_r \end{pmatrix} R$$

where R is upper-triangular. That is, we have the *QR factorization of A*:

$$A = QR$$

where Q is an n by r matrix whose columns are orthonormal and R is an invertible r by r upper-triangular matrix.

Example: a) Find an orthonormal basis for the subspace of $\mathbf{R}^4$ spanned by $(1, 0, 0, -1), (0, 0, 1, 2), (1, -1, 0, 2)$.

b) Find the QR factorization of $A = \begin{pmatrix} 1 & 0 & 1 \\ 0 & 0 & -1 \\ 0 & 1 & 0 \\ -1 & 2 & 2 \end{pmatrix}$.

Each part is accomplished in a single step.

```
Command:  get 137 a
Get matrix 137 and store it in A

     1.        0.        1.
     0.        0.       -1.
     0.        1.        0.
    -1.        2.        2.

Command:  rank a
Rank -- record the rank of A

The following upper-triangular matrix has the same rank as A:

    -2.45     -1.63      0.41
              -1.53      0.87
                        -1.04

Enter the rank of A (it is at most 3):  3

Command:  orthonormalize a
Orthonormalize the columns of A and store the result in %

    -0.71     -0.58      0.32
     0.        0.       -0.63
     0.       -0.58     -0.63
     0.71     -0.58      0.32
```

The columns of this matrix form the desired orthonormal basis.

```
Command:  qrfactor a
QRfactor A and store the Q and R factors in [% %%]

The product of the following Q and R factors is A.

    -0.71     -0.58      0.32
     0.        0.       -0.63
     0.       -0.58     -0.63
     0.71     -0.58      0.32

    -1.41      1.41      0.71
     0.       -1.73     -1.73
     0.        0.        1.58
```

Note that the first matrix (the Q factor) is just the matrix obtained in response to the **orthonormalize** command, as it should be.

Remark: For reasons of numerical stability and speed of computation, MAX uses a different algorithm from Gram-Schmidt to produce orthonormal vectors. See Appendix II.

PROBLEMS

1. For each of the following collections of vectors $v_1, \ldots, v_r$ in $\mathbf{R}^4$, find an orthonormal basis for the subspace of $\mathbf{R}^4$ spanned by $v_1, \ldots, v_r$.

   a) [138] $(-3, 4, 0, 0), (0, 1, 2, -2)$

   b) [3] $(5, 2, -1, 1), (0, 1, -1, 1), (3, -7, 1, 2)$

   c) [139] $(-3, 1, 1, 2), (0, 4, 1, 5), (1, 0, 0, 0)$

   d) [140] $(-3, 1, 1, 2), (-1, 3, 1, 4), (0, 4, 1, 5), (1, 0, 0, 0)$

   e) [141] $(-3, 1, 1, 2), (1, 1, 0, 1), (-1, 3, 1, 4), (0, 4, 1, 5), (1, 0, 0, 0)$

   (What is different about parts d) and e) from the earlier parts and from the discussion in this section?)

2. Find an orthonormal basis for each of the following planes in $\mathbf{R}^3$.

   a) $x - y = 0$

   b) $2x + y - 3z = 0$

3. [54, 141, 128] For each of the indicated matrices A, do the following parts.

   a) Find an orthonormal basis for the nullspace of A.

   b) Find an orthonormal basis for the row space of A.

   c) Check that the bases in a) and b) taken together form an orthonormal basis for $\mathbf{R}^5$.

4. [138, 3, 142] For each of the indicated matrices A, find the QR factorization of A.

Section 2.6 LEAST-SQUARES SOLUTIONS OF $AX = K$

Let A be an n by r matrix of rank r, K be a column n-vector. The equation $AX = K$ has solutions iff K lies in the column space of A. (Recall that $AX = A_1 x_1 + \ldots + A_r x_r$ if $A = \begin{pmatrix} A_1 \ldots A_r \end{pmatrix}$ and $X = \begin{pmatrix} x_i \end{pmatrix}$.) The *least-squares solution* of $AX = K$ is the vector $\bar{X}$ so that $K - A\bar{X}$ is orthogonal to the column space of A (as in the figure).

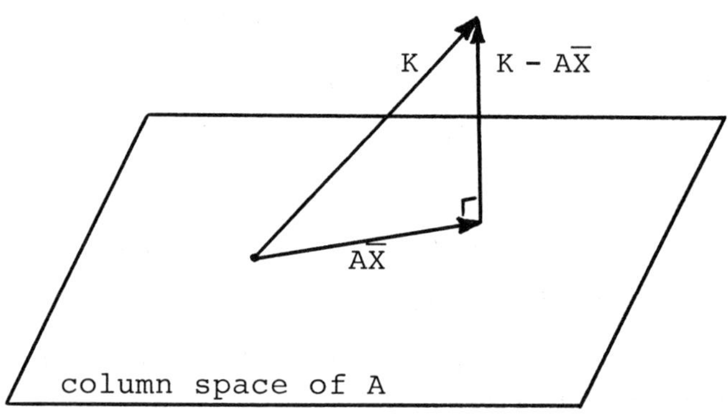

Hence $\bar{X}$ is the vector so that

$$(AY)'(K - A\bar{X}) = 0 \quad \text{for all } Y,$$

$$\text{i.e.,} \quad Y'(A'K - A'A\bar{X}) = 0 \quad \text{for all } Y,$$

$$\text{i.e.,} \quad A'K - A'A\bar{X} = 0,$$

$$\text{i.e.,} \quad A'A\bar{X} = A'K.$$

The equation $A'AX = A'K$ is known as the *normal equation*; its solution, we have just seen, is the least-squares solution $\bar{X}$ of $AX = K$. (Note: The phrase "least-squares solution" comes from the fact that the length $\| K - AX \|$ is smallest when $X = \bar{X}$ and from the fact that the length of a vector is the square root of the sum of the squares of its components.)

Since the columns of A are assumed to be linearly independent, the matrix $A'A$ is invertible. Hence we can solve for $\bar{X}$ and get

$$\bar{X} = (A'A)^{-1} A'K.$$

The QR factorization of A simplifies this expression. For if $A = QR$,

$$\begin{aligned}
\bar{X} &= ((QR)'QR)^{-1}(QR)'K \\
&= (R'Q'QR)^{-1}R'Q'K \\
&= (R'R)^{-1}R'Q'K \qquad (Q'Q = I) \\
&= R^{-1}(R')^{-1}R'Q'K \\
&= R^{-1}Q'K.
\end{aligned}$$

That is, given $AX = K$, to find $\bar{X}$ we form $Q'K$ and solve the upper triangular system $RX = Q'K$ by back-substitution. This is roughly what MAX does to solve $AX = K$ whenever A is singular. (Recall from Section 1.3 that the LU factorization is used if A is nonsingular.) Appendix II contains more details.

Example: Let A and K be matrices 143 and 144 in group storage. Find the least-squares solution of $AX = K$.

```
Command:  get 143 a
Get matrix 143 and store it in A
       0.         1.         1.
       3.         0.        -1.
      -2.         2.         0.
       4.         1.         0.
Command:  get 144 k
Get matrix 144 and store it in K
      -1.
       4.
      -5.
       1.
Command:  rank a
Rank -- record the rank of A

The following upper-triangular matrix has the same rank as A:
   -5.39        0.        0.56
                2.45      0.41
                          1.23

Enter the rank of A (it is at most 3):  3
```

```
Command:  solve a k xbar
Solve the system A K and store the solution in XBAR
     0.85
    -1.6
    -0.42
WARNING: This is a least-squares solution, which need not be an
         actual solution
Command:  compute k - a*xbar
Compute (K - (A*XBAR)) and store in %
     1.02
     1.02
    -0.1
    -0.82
```

The matrix $K - A\bar{X}$ is the *matrix of residuals*; it gives a measure of the error involved in the least-squares solution of $AX = K$. (Recall the figure at the beginning of this section.) The vector $K - A\bar{X}$ is zero iff K lies in the column space of A, i.e., iff $\bar{X}$ is an actual solution of $AX = K$. In this example, the least-squares solution is not an actual solution. We can use the QR factorization to check our answer $\bar{X}$.

```
Command:  qrfactor a [q r]
QRfactor A and store the Q and R factors in [Q R]
The product of the following Q and R factors is A.
       0.          0.41         0.68
      -0.56        0.          -0.56
       0.37        0.82        -0.44
      -0.74        0.41         0.2

      -5.39        0.           0.56
       0.          2.45         0.41
       0.          0.           1.23
Command:  compute r^-1*q'*k
Compute (((R^-1)*Q')*K) and store in %
     0.85
    -1.6
    -0.42
```

PROBLEMS

1. [35 & 36, 145 & 146, 147 & 148] For each of the indicated pairs of matrices A & K, do the following parts.

   a) Find the least-squares solution of $AX = K$.

   b) Find the matrix of residuals.

   c) Check your answer in part a) using the QR factorization of A.

2. [7 & 44, 149 & 150] Each matrix A in Problem 1 has linearly independent columns. MAX produces a least-squares solution of $AX = K$ for any matrix A. It does this by solving $BX = K$, where B is a matrix consisting of the linearly independent columns of A, and then inserting 0's into the solution in positions corresponding to the linearly dependent columns of A.

   a) For each of the indicated pairs of matrices A & K, use the **solve** command to find a least-squares solution of $AX = K$.

   b) Note the 0's in the solutions in part a). Use the **cbasis** command to confirm that 0's occur in positions corresponding to linearly dependent columns of A.

Section 2.7 LEAST-SQUARES APPLICATIONS

The data in a wide variety of applications problems, when plotted on appropriate coordinate axes, appear to lie along a straight line. Suppose we want to fit the straight line $y = a + bx$ to the n data points $(x_1, y_1), \ldots, (x_n, y_n)$. We attempt to find a and b so that

$$y_1 = a + bx_1$$
$$\vdots$$
$$y_n = a + bx_n$$

or $Y = AX$ where $Y = \begin{pmatrix} y_1 \\ \vdots \\ y_n \end{pmatrix}, A = \begin{pmatrix} 1 & x_1 \\ \vdots & \vdots \\ 1 & x_n \end{pmatrix}, X = \begin{pmatrix} a \\ b \end{pmatrix}$. The least-squares solution of $AX = Y$ gives the line that best fits the n points. The resulting equation $y = a + bx$ is called the *regression equation*.

Similarly, if z depends linearly on x and y, we can use the method of least-squares to find a, b, c so that $z = a + bx + cy$ (again called the *regression equation*) best fits n data points $(x_1, y_1, z_1), \ldots, (x_n, y_n, z_n)$. Further generalizations are possible, as seen in some of the following examples and problems.

Example 1: Suppose the average weights of men vary with their heights according to the following table.

Height	Weight
5'4"	141 lb
5'6"	148 lb
5'8"	155 lb
5'10"	163 lb
6'	172 lb
6'2"	182 lb
6'4"	190 lb
6'6"	198 lb

Fit a least-squares straight line to the data. Estimate the average weights of men of heights 5'3", 5'11", 6'3", 6'10".

We convert all heights to inches and solve $AX = Y$ for the appropriate A and Y.

Command: **get 152 b**
Get matrix 152 and store it in B

64.	141.
66.	148.
68.	155.
70.	163.
72.	172.
74.	182.
76.	190.
78.	198.

Command: **ones**

Ones -- store the Integer1 by Integer2 matrix of ones in Matrix1
Integer1: **8**
Integer2: **1**
Matrix1: <%>
Ones -- store the 8 by 1 matrix of ones in %

```
1.
1.
1.
1.
1.
1.
1.
1.
```

Command: **aug % b m**
Augment % by B and store the result in M

1.	64.	141.
1.	66.	148.
1.	68.	155.
1.	70.	163.
1.	72.	172.
1.	74.	182.
1.	76.	190.
1.	78.	198.

```
Command: solve m(,1-2) m(,3)

Enter the rank of M(1-8,1-2) (it is at most 2): 2
Solve the system M(1-8,1-2) M(1-8,3-3) and store the solution in %

   -126.79
      4.16

WARNING: This is a least-squares solution, which need not be an
         actual solution
```

This says the best straight line giving weight y in terms of height x is $y = -126.79 + 4.16x$. To find the average weights of men of the given heights, we evaluate y for $x = 63, 71, 75, 82$. We get average weights of 135 lb, 169 lb, 185 lb, 214 lb for men of heights 5'3", 5'11", 6'3", 6'10".

Example 2: A chemistry experiment is designed to analyze the dependence of vapor pressure of water (P, in mm of Hg) on temperature (T, in degrees Kelvin). The relationship between P and T is given by the Clausius-Clapeyron equation

$$ln(P) = (-\Delta H_v/R)(T^{-1}) + b,$$

a linear relation between the variables $ln(P)$ and T^{-1}. By fitting a straight line to data points given by values of P and T, we can compute the best heat of vaporization (ΔH_v). (R is the known constant 8.31 J/K mol.)

Suppose an experiment yields the following data:

temperature t (in $^\circ C$)	pressure P (in mm of Hg)
0	4.58
10	9.21
20	17.54
30	31.82
40	55.3
50	92.5

Convert the data and place it in a 6 by 3 matrix M (number 153); the first column is all 1's, the second column is $T^{-1} = 1/(t + 273.15)$, and the third column is $ln(P)$. Now solve $AX = Y$ where A is the first two columns of M and Y is the third column of M. The least-squares solution is $\begin{pmatrix} 20.9447 \\ -5299.89 \end{pmatrix}$.

Hence the best straight-line fit to the data is

$$ln(P) = (-5299.89)(T^{-1}) + 20.9447.$$

Thus we compute the best heat of vaporization to be

$$\Delta H_v = (5299.89)(8.31) = 44042 \text{ J/mol}.$$

PROBLEMS

1. [154] During the 1970's, Hong Kong experienced a yearly decline in its birth rate as shown in the following table. (Source: <u>Demographic Yearbook</u> 1981.)

Year	Birth rate (number of births/1000 females)
1972	86.6
1973	85.6
1974	85.3
1975	77.3
1976	73.7
1977	72.0
1978	70.3
1979	66.2
1980	66.1

a) Find the best straight line fit to the data.

b) Assume the decline continues into the 1980's. Use your answer in part a) to predict the year in which the birth rate drops to half of what it was in 1972.

2. [155] Consider the following data for a certain sample of trees.

Diameter D (at chest height, in inches)	Height H (in feet)	Volume V (of usable wood, in cubic feet)
14.5	76	31.4
14.2	75	29.9
15.7	85	40.8
14.4	76	31.0
16.8	71	35.5
20.5	82	65.8
14.1	80	32.7
14.8	69	31.3
18.9	72	48.4
14.0	66	25.5

We are interested in predicting the volume of wood that will come from a tree with certain measurements. It is easy to find the diameter but may be rather difficult to find its height.

a) Find the regression equation that gives V in terms of D.
For each 1" increase in diameter, what is the predicted increase in volume?

b) Find the regression equation that gives V in terms of D and H.
Find the difference in volume between two trees of the same height whose diameter differ by 1".

c) If a tree has diameter 17.6" and height 81', predict its volume of wood two ways, using a) and b). Which number is likely to be more accurate?

3. [156] Consider the following data of disposable personal income Y_t and personal consumption expenditures C_t. (Source: <u>Economic Report of the President</u> January, 1987.)

Year	Personal Income Y_t (in billions of 1982 dollars)	Consumption Expenditures C_t (in billions of 1982 dollars)
1966	1431.3	1298.9
1967	1493.2	1337.7
1968	1551.3	1405.9
1969	1599.8	1456.7
1970	1668.1	1492.0
1971	1728.4	1538.8
1972	1797.4	1621.9
1973	1916.3	1689.6
1974	1896.6	1674.0
1975	1931.7	1711.9
1976	2001.0	1803.9
1977	2066.6	1883.8
1978	2167.4	1961.0
1979	2212.6	2004.4
1980	2214.3	2000.4
1981	2248.6	2024.2
1982	2261.5	2050.7
1983	2331.9	2146.0
1984	2470.6	2246.3
1985	2528.0	2324.5

a) Find the regression equation that gives C_t in terms of Y_t. Each dollar of tax cut increases disposable personal income by \$1. With this in mind, use the regression equation to predict the increase in personal consumption expenditures that would result from a tax cut of \$10 billion.

b) Often an increase in consumer spending lags behind an increase in income. This can be taken into account by considering C_t as a linear function of Y_t and C_{t-1} (i.e., expenditures in a given year depend linearly on income in that year and expenditures in the previous year). Find the regression equation that gives C_t in terms of Y_t and C_{t-1}. (Start with year 1967.) Suppose a tax cut of \$10 is proposed to stimulate consumer spending which has become constant. Use the regression equation to predict the resulting increase in personal consumption expenditures.

4. [157] Consider the following data for hydrolysis of sucrose in an aqueous solution.

Time t (in minutes)	Sucrose S (percent remaining)
10	96.1
60	80.0
95	69.8
145	57.9
295	32.2
590	11.0

Assume this is a first-order reaction so $S = 100e^{-kt}$. Use least-squares to find the rate constant k. (Take the natural logarithm. The third column in matrix 157 contains values of $ln(S)$.)

5. [158] The following equation is used in a chemistry experiment to analyze the absorption spectrum of molecular iodine I_2:

$$\sigma = \sigma_e + w'_e(v' + \frac{1}{2}) - w'_e x'_e(v' + \frac{1}{2})^2$$
$$- w''_e(v'' + \frac{1}{2}) + w''_e x''_e(v'' + \frac{1}{2})^2.$$

Here v' and v'' are vibrational quantum numbers for the observed frequency σ, and the remaining quantitites are parameters that are based on a theoretical model. The following data were obtained in a student experiment.

v'	v''	σ (in cm^{-1})
14	0	17306.4
15	0	17408.0
16	0	17506.7
17	0	17603.8
13	1	16991.1
14	1	17092.5
15	1	17193.9
16	1	17290.9
10	2	16463.4
11	2	16566.0
12	2	16672.9
13	2	16779.5

a) Find the regression equation that gives best values for the five parameters $\sigma_e, w'_e, w'_e x'_e, w''_e, w''_e x''_e$.
(Compare your answers with the following accepted values:
$w'_e = 125.273, w'_e x'_e = .7016, w''_e = 214.52$.)

b) Predict the value of σ when $v' = 1$ and $v'' = 3$.
(The value given in one extensive table is 15243.4.)

Chapter 3

Eigenvalues and Eigenvectors

Section 3.1 EIGENVALUES AND EIGENVECTORS OF A

Given an n by n matrix A, we consider a scalar λ and a nonzero vector X so that $AX = \lambda X$. Such a λ is an *eigenvalue* of A and X is an *eigenvector of A corresponding to* λ.

Note that λ is an eigenvalue of A iff $(\lambda I - A)X = 0$ for some nonzero X. Hence the nullspace of the matrix $\lambda I - A$ consists of the eigenvectors of A corresponding to λ (plus the zero vector) and is the *eigenspace of A corresponding to* λ.

If A has n linearly independent eigenvectors, then A can be diagonalized: $P^{-1}AP = D$ where P is the invertible matrix whose columns are the eigenvectors $X_1, \ldots, X_n$ and D is the diagonal matrix whose diagonal entries are the eigenvalues $\lambda_1, \ldots, \lambda_n$.

(Proof:

$$AP = A\left(X_1 \ldots X_n\right) = \left(AX_1 \ldots AX_n\right) = \left(\lambda_1 X_1 \ldots \lambda_n X_n\right)$$

$$= \left(X_1 \ldots X_n\right) \begin{pmatrix} \lambda_1 & & 0 \\ & \ddots & \\ 0 & & \lambda_n \end{pmatrix} = PD. \;)$$

Example: Let A and B be matrices 1 and 23 in group storage and find their eigenvalues and eigenvectors.

```
Command:  get 1 a
Get matrix 1 and store it in A
     -5.          1.          5.
     -7.          4.          4.
     -1.          1.          1.
Command:  eig a
Eigenvalues and eigenvectors of A -- store in [% %%]
Each eigenvector is displayed below its eigenvalue.
     -4.          3.          1.

      1.          0.33        1.
      0.86        1.          1.
   2.78E-02       0.33        1.

Command:  get 23 b
Get matrix 23 and store it in B
      5.          0.          2.
      0.          1.          0.
     -4.          0.         -1.
Command:  eig b
Eigenvalues and eigenvectors of B -- store in [% %%]
Each eigenvector is displayed below its eigenvalue.
      3.          1.          1.

      1.         -0.5         0.
      0.          0.          1.
     -1.          1.          0.
```

Matrix A has three distinct eigenvalues and hence three linearly independent eigenvectors. Matrix B has repeated eigenvalue 1 but still has a full complement of three linearly independent eigenvectors. Both A and B can be diagonalized. Note that eigenvectors are scaled so that the component with largest absolute value is 1.

PROBLEMS

1. [8, 161, 162, 163] For the indicated matrices A, do the following parts.

   a) Find the eigenvalues and eigenvectors of A.

   b) If A can be diagonalized, give a nonsingular matrix P and a diagonal matrix D so that $P^{-1}AP = D$.

   Check using the **compute** command.

2. [142, 164] For the indicated matrices A, find the eigenvalues and eigenvectors of A.

   WARNING: The algorithm in MAX assumes that an n by n matrix will have n linearly independent eigenvectors. When this is not true, as in this problem, the results can be hard to interpret, especially since they are obscured by larger than usual round-off errors. See the discussion in Appendix II.

3. For each matrix A given below, use the **nbasis** command to find the eigenspace of A corresponding to the given eigenvalues. Check that your answers agree with those in Problem 1.

       Matrix 8 has eigenvalues 1, 2, 3.
       Matrix 162 has eigenvalues 3, 6.

4. [8, 162, 142] For the indicated matrices A, do the following parts.

   a) Find the eigenvalues of A, A', A^{-1}, A^2.

   b) In general, how are the eigenvalues of A and A' related?
   How are the eigenvalues of A and A^{-1} related?
   How are the eigenvalues of A and A^2 related?
   How are the eigenvalues of A and A^n related?

   c) Prove your assertions in part b).

5. [9] The indicated matrix A is idempotent, i.e., $A^2 = A$.

   a) Find the eigenvalues of A.

   b) In general, what can you say about the eigenvalues of an idempotent matrix?

   c). Prove your assertion in part b).

6. [14, 15] The indicated matrices A are nilpotent, i.e., $A^n = 0$ for some integer n.

   a) Find the eigenvalues of A.

b) In general, what can you say about the eigenvalues of a nilpotent matrix?
(See the **WARNING** message in problem 2.)

c) Prove your assertion in part b).

7. [16, 24] Matrix 16 satisfies $A^3 - 12A + 16I = 0$. Matrix 24 satisfies $A^2 - 5A + 4I = 0$.

   a) Find the eigenvalues of each matrix A.

   b) In general, what can you say about the eigenvalues of a matrix that satisfies a polynomial equation?

   c) Prove your assertion in part b).

8. [1] For the indicated matrix A, let $B_k = A - kI$ for $k = 1, 2, 3$.

   a) Find the eigenvalues and eigenvectors of B_k.

   b) In general, how are the eigenvalues and eigenvectors of matrices A and $A - kI$ related?

   c) Prove your assertion in part b).

9. [1 & 165] For the indicated pair of matrices A & B, do the following parts.

   a) Show that $AB = BA$.

   b) Find the eigenvectors of A and B.

   c) What conjecture might you make based on parts a) and b)? (Assume matrices A and B have all distinct eigenvalues.)

   d) Prove your assertion in part c).

The last two problems deal with the *adjacency matrix* of a graph, discussed at the end of Section 1.5.

10. Let K_n be the *complete graph* on n vertices; in this graph, each pair of distinct vertices is joined by an edge. The figure shows K_5.

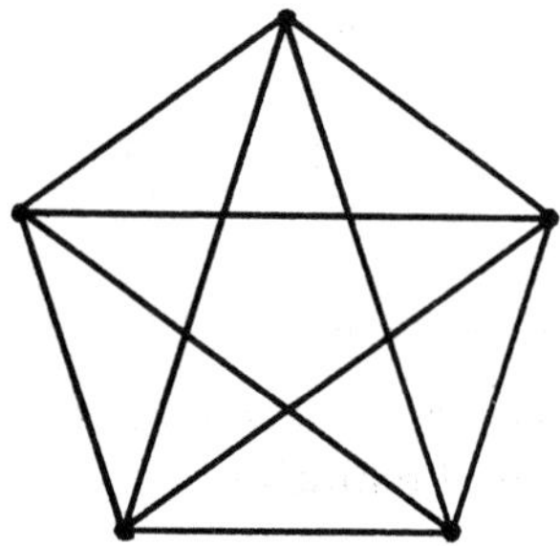

    a) Find the adjacency matrix A_n for K_n.

    b) Find the eigenvalues of A_4, A_5, A_6.
       (You can use the **ones** command to help construct the matrices.)

    c) What conjecture might you make about the eigenvalues of A_n?

11. Let C_n be the *cycle* on n vertices. The figure shows C_5.

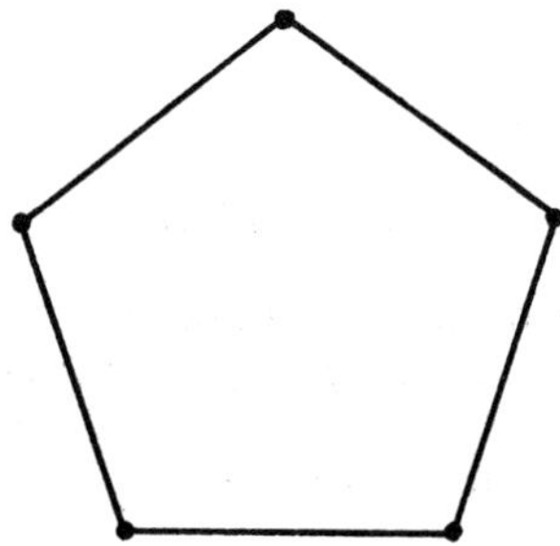

    a) Find the adjacency matrix A_n for C_n.

    b) Find the eigenvalues of A_4, A_5, A_6.
       Express each eigenvalue λ_j in terms of $\cos \theta_j$ where θ_j is an angle in radians.

    c) What conjecture might you make about the eigenvalues of A_n?

Section 3.2   DIFFERENCE EQUATIONS APPLICATIONS

We consider situations which give rise to a sequence of column n-vectors $Y_0, Y_1, Y_2, \ldots$ that satisfy $Y_k = AY_{k-1}$ for $k = 1, 2, \ldots$, where A is a fixed n by n matrix. It follows that $Y_k = A^k Y_0$ for any such k.

Suppose A has n linearly independent eigenvectors $X_1, \ldots, X_n$ corresponding to eigenvalues $\lambda_1, \ldots, \lambda_n$. Then there exist constants $c_1, \ldots, c_n$ such that

$$Y_k = c_1 \lambda_1^k X_1 + \ldots + c_n \lambda_n^k X_n.$$

(Proof: Given Y_0 (the initial conditions), there exist $c_1, \ldots, c_n$ so that

$c_1 X_1 + \ldots + c_n X_n = Y_0$

or $PC = Y_0$ where

$$P = \begin{pmatrix} X_1 \ldots X_n \end{pmatrix}, \quad C = \begin{pmatrix} c_1 \\ \vdots \\ c_n \end{pmatrix}. \quad \text{Then}$$

$$\begin{aligned} Y_k = A^k Y_0 &= A^k(c_1 X_1 + \ldots + c_n X_n) \\ &= c_1(A^k X_1) + \ldots + c_n(A^k X_n) \\ &= c_1 \lambda_1^k X_1 + \ldots + c_n \lambda_n^k X_n.) \end{aligned}$$

We can use MAX to find the λ_i, X_i, c_i. The c_i are computed from $PC = Y_0$; note that P is exactly the matrix that MAX puts the eigenvectors into.

PROBLEMS

Markov Chains

Suppose A is a *stochastic* (or *Markov* or *probability*) matrix, i.e., each entry of A is nonnegative and the sum of the entries in any column of A is 1. Then the process described above is a *Markov process* or *Markov chain* and A is a *transition* matrix.

It can be shown that 1 is a *dominant* eigenvalue of A, i.e., $|\lambda_j| \leq 1$ for all j. Further, it often happens that one eigenvalue, say λ_1, is 1 and all other eigenvalues λ_j satisfy $|\lambda_j| < 1$. Then 1 is a *strictly dominant* eigenvalue of A, and it follows that $Y_k \to c_1 X_1$ as $k \to \infty$ (because $\lambda_j^k \to 0$ for $j > 1$).

If the sum of the coordinates of each Y_k is 1, then Y_k approaches that eigenvector corresponding to eigenvalue 1 whose coordinates sum to 1; this limit is the *steady state* of the Markov process. In MAX, the **scale** command is used to multiply a column by a scalar so that the column sums to 1. Thus, rather than finding c_1, we can simply scale X_1 and the result will be the limit of Y_k.

1. [166] Companies 1, 2, 3 each produce the same product. Suppose that each year:

   Company 1 retains 85% of its customers, loses 5% to 2 and 10% to 3,
   Company 2 retains 75% of its customers, loses 15% to 1 and 10% to 3,
   Company 3 retains 90% of its customers, loses 5% to 1 and 5% to 2.
   Let Y_k be the column 3-vector that gives the share of the market for the three companies at the end of year k.

   a) Derive the 3 by 3 matrix A so that $Y_k = AY_{k-1}$.
   Compare your answer with matrix 166.

   b) Suppose the initial share is $Y_0 = \begin{pmatrix} .3 \\ .5 \\ .2 \end{pmatrix}$.
   Find $Y_1, Y_2, Y_3, Y_4, Y_5, Y_{10}, Y_{20}$.

   c) Find the eventual share of the market held by each of the three companies. Does this depend on the initial share Y_0?

2. [167] Consider a chicken trying to negotiate a 4-lane highway. The chicken can be in six possible states ranging from $1 = $ left shoulder of the highway to $6 = $ right shoulder of the highway (see the figure).

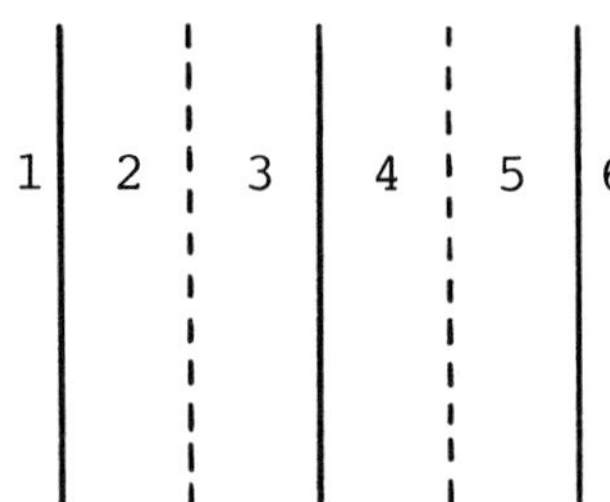

We suppose the following: If the chicken is in either of states 1 and 6, it stays put or moves to the neighboring state with equal probability; if the chicken is in any of states 2 through 5, it stays put or moves to either neighboring state with equal probability.

(All movements occur in a given fixed time period, say 10 seconds.)
Let Y_k be the column 6-vector of probabilities that the chicken is in states $1, \ldots, 6$ after k time periods.

a) Derive the 6 by 6 matrix A so that $Y_k = AY_{k-1}$.
Compare your answer with matrix 167.

b) If the chicken starts on the left shoulder of the highway, find the proportion of time that the chicken spends in each of the 6 states over the long run.

3. [168] The figure shows the corridors of an art gallery with intersections labeled 1 through 7. A guard is to patrol the corridors, remaining at an intersection for 15 minutes and then moving to a neighboring intersection. The new intersection is chosen randomly, all possible choices being equally likely. After any break (including overnight and days off), the guard starts at a neighbor of the intersection last occupied.

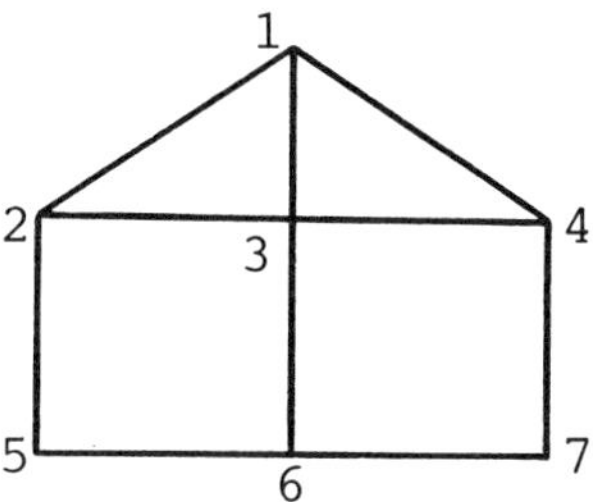

Let Y_k be the column 7-vector of probabilities that the guard is at intersections $1, \ldots, 7$ after k 15-minute time periods.

a) Derive the 7 by 7 matrix A so that $Y_k = AY_{k-1}$.
Compare your answer with matrix 168.

b) Over the long run, what proportion of time does the guard spend at each intersection?

c) Redo parts a) and b) under the assumption that at the end of each 15-minute time period, the guard either moves to a neighboring intersection or stays put. The (possibly) new intersection is chosen randomly with all possible choices being equally likely.

4. [169] A high-stakes gambler plays successive rounds of a game of chance in which on each round his probability of winning $1000 is 0.35, of

staying the same is 0.25, of losing \$1000 is 0.40. He stops if he loses all his money (state 0) or if his earnings reach \$5000 (state 5).

(This is an example of a Markov process with *absorbing states* in contrast to Problem 2 which had *reflecting states*.)

Let Y_k be the column 6-vector of probabilities that the gambler is in states $0, \ldots, 5$ after k rounds of the game.

   a) Derive the 6 by 6 matrix A so that $Y_k = AY_{k-1}$.
      Compare your answer with matrix 169.

After playing a very long time, the gambler is certain to have stopped, either having gone broke or having won; i.e., as $k \to \infty$,

$$Y_k \to \begin{pmatrix} p \\ 0 \\ 0 \\ 0 \\ 0 \\ 1-p \end{pmatrix}.$$

   b) Find p if the gambler starts with \$2000; with \$3000.

   c) Extend part b) by finding p if the gambler starts with \$0, \$1000, \$2000, \$3000, \$4000, \$5000. (This can be done all at once. How?)

   d) How does this problem differ from the previous three problems?

Age-Specific Populations

Consider an example of an animal population whose size is determined by characteristics of the female. (The assumption is that there will always be enough males to propogate the species.) Suppose the maximal lifespan of the females is three years. On the average, 37% of the newborns live to be one year old and 30% of the one-year olds live to be two years old. Further, on the average, each one-year old produces 3 female offspring and each two-year old produces 1 female offspring. Let Y_k be the column 3-vector that gives the number of females in the three age groups in year k. Then $Y_k = AY_{k-1}$ where $A = \begin{pmatrix} 0 & 3 & 1 \\ .37 & 0 & 0 \\ 0 & .30 & 0 \end{pmatrix}$.

Such a population model is called a *Leslie model* and A is a *Leslie matrix*. In general,

$$
A = \begin{pmatrix}
a_1 & a_2 & a_3 & \cdots & a_{n-1} & a_n \\
p_1 & 0 & 0 & \cdots & 0 & 0 \\
0 & p_2 & 0 & \cdots & 0 & 0 \\
& & \vdots & & & \\
0 & 0 & 0 & \cdots & p_{n-1} & 0
\end{pmatrix}
$$

where a_i is the number of offspring per individual in group i and p_i is the probability that an individual in group i survives one period to become a member of group $i+1$.

It can be shown that a Leslie matrix has a unique positive eigenvalue, say λ_1, that is simple (not repeated) and dominant, i.e., $|\lambda_j| \le \lambda_1$ for $j > 1$. Further, under mild conditions on the matrix, λ_1 is strictly dominant, i.e., $|\lambda_j| < \lambda_1$ for $j > 1$. In this case, $Y_k \approx c_1 \lambda_1^k X_1$ and $Y_k \approx \lambda_1 Y_{k-1}$ for large k, where X_1 is an eigenvector corresponding to λ_1. (To see the first assertion, divide both sides of the equation $Y_k = c_1 \lambda_1^k X_1 + \ldots + c_n \lambda_n^k X_n$ by λ_1^k and let k get large; the second assertion follows from the first.) Thus λ_1 gives the long-term growth rate per time period, and from X_1 we get the long-term proportions of the population in the different age groups.

5. [40] Consider the animal population in the above example. Suppose initially there are 100 females in each of the three age groups.

   a) Find the number of females in each age group after 1, 2, 3, 4, 5, 10, 20 years.

   b) Find the long-term yearly growth rate and the long-term proportion of females in each age group.

6. [170] Consider an animal population whose females have the following characteristics. The maximal lifespan is four years. The probability that a newborn lives to be one year old is $\frac{5}{13}$; the probability a one-year old lives to be two is $\frac{3}{10}$; the probability a two-year old lives to be three is $\frac{1}{4}$. On the average, each one-year old produces 2 female offspring; each two-year old produces 2 more female offspring.

   a) Derive the 4 by 4 Leslie matrix.
      Compare your answer with matrix 170.

   b) Find the long-term yearly growth rate and the long-term proportion of females in each age group.

7. The tables below give birth and death parameters (the a_i and p_i in a
   Leslie matrix) for three populations of females. (Source from which the
   data was derived: <u>Demographic Yearbook</u> 1981.)

Venezuela (1979)

Age interval (in years)	a_i	p_i
0-10	0	.953
10-20	.259	.994
20-30	1.099	.990
30-40	.699	.983
40-50	.154	.963
50-60	0	.915
60-70	0	---

Netherlands (1979)

Age interval (in years)	a_i	p_i
0-10	0	.990
10-20	.022	.997
20-30	.532	.996
30-40	.209	.993
40-50	.009	.980
50-60	0	.952
60-70	0	---

Bulgaria (1980)

Age interval (in years)	a_i	p_i
0-10	0	.978
10-20	.196	.996
20-30	.687	.994
30-40	.107	.990
40-50	.006	.976
50-60	0	.938
60-70	0	---

a) [171] Find the long-term growth rate per time period (10 years) for Venezuela and the eventual distribution of females in each age category.

b) [172] Repeat part a) for the Netherlands.

c) [173] Repeat part a) for Bulgaria.

d) Describe the differences between the eventual distributions of females in the different age categories for populations that are increasing, decreasing, stable.

Genetics

Consider X-linked inheritance; here the male of the species possesses only one of two possible genes (denoted A and a) and the female possesses a pair of the two genes (denoted AA, Aa and aa). A male offspring receives one of his mother's two genes with equal probability, and a female offspring receives the one gene of her father and one of her mother's two genes with equal probability. This gives the following table of genotype probabilities:

Genotype of offspring	Genotypes of Parents (father, mother)					
	(A,AA)	(A,Aa)	(A,aa)	(a,AA)	(a,Aa)	(a,aa)
A	1	1/2	0	1	1/2	0
a	0	1/2	1	0	1/2	1
AA	1	1/2	0	0	0	0
Aa	0	1/2	1	1	1/2	0
aa	0	0	0	0	1/2	1

Consider a program of inbreeding where we begin initially with a male and female, select one offspring of each sex at random and mate them, and continue the process. Let Y_k be the column 6-vector of probabilities that the sibling pair in the kth generation is of type (A,AA), (A,Aa), (A,aa), (a,AA), (a,Aa), (a,aa).

8. [174] a) Derive the 6 by 6 matrix B so that $Y_k = BY_{k-1}$. Compare your answer with matrix 174.

b) Use the eigenvalues and eigenvectors of B to give a general expression for $\lim_{k \to \infty} Y_k$ for an arbitrary Y_0.

c) If $Y_0 = \begin{pmatrix} 0 \\ 0 \\ 1 \\ 0 \\ 0 \\ 0 \end{pmatrix}$, i.e., the initial parents are (A,aa), find $\lim_{k \to \infty} Y_k$.

d) If $Y_0 = \begin{pmatrix} 0 \\ 0 \\ 0 \\ 1 \\ 0 \\ 0 \end{pmatrix}$, i.e., the initial parents are (a,AA), find $\lim_{k \to \infty} Y_k$.

e) If the initial parents are equally likely to be any of the six possible genotypes, i.e.,

$$Y_0 = \begin{pmatrix} 1/6 \\ 1/6 \\ 1/6 \\ 1/6 \\ 1/6 \\ 1/6 \end{pmatrix},$$

find $\lim_{k \to \infty} Y_k$.

f) The mathematics involved in this problem is similar to that in a previous problem. Which one?

Section 3.3 DIFFERENTIAL EQUATIONS APPLICATIONS

A system of n first-order linear differential equations

$$\frac{dy_1}{dt} = a_{11}y_1 + \ldots + a_{1n}y_n$$

$$\vdots$$

$$\frac{dy_n}{dt} = a_{n1}y_1 + \ldots + a_{nn}y_n$$

can be written in matrix form as $\frac{dY}{dt} = AY$ where $A = \left(a_{ij}\right)$ and $Y = \left(y_j\right)$. If the n by n matrix A has n linearly independent eigenvectors $X_1, \ldots, X_n$ corresponding to eigenvalues $\lambda_1, \ldots, \lambda_n$, the general solution of $\frac{dY}{dt} = AY$ is

$$Y = c_1 e^{\lambda_1 t} X_1 + \ldots + c_n e^{\lambda_n t} X_n$$

where $c_1, \ldots, c_n$ are arbitrary constants. (This is derived as follows: Let

$$P = \left(X_1 \ldots X_n \right), \quad D = \begin{pmatrix} \lambda_1 & & 0 \\ & \ddots & \\ 0 & & \lambda_n \end{pmatrix}.$$

The substitution $Y = PZ$ reduces the system $\frac{dY}{dt} = AY$ to the uncoupled system $\frac{dZ}{dt} = DZ$ whose general solution is

$$Z = \begin{pmatrix} c_1 e^{\lambda_1 t} \\ \vdots \\ c_n e^{\lambda_n t} \end{pmatrix}.$$

Then $Y = PZ = \left(X_1 \ldots X_n \right) \begin{pmatrix} c_1 e^{\lambda_1 t} \\ \vdots \\ c_n e^{\lambda_n t} \end{pmatrix} = c_1 e^{\lambda_1 t} X_1 + \ldots + c_n e^{\lambda_n t} X_n.)$

To solve an initial-value problem

$$\frac{dY}{dt} = AY, \quad Y(0) = Y_0,$$

we find the constants c_i by solving the system of linear equations $PC = Y_0$

for the column n-vector $C = \begin{pmatrix} c_1 \\ \vdots \\ c_n \end{pmatrix}$.

PROBLEMS

1. Three tanks, each holding 100 gallons of liquid, are interconnected as shown in the figure. The numbers represent flow rates (in gal/min) with arrows indicating the direction of the flow. Pure water enters the system at tank I.

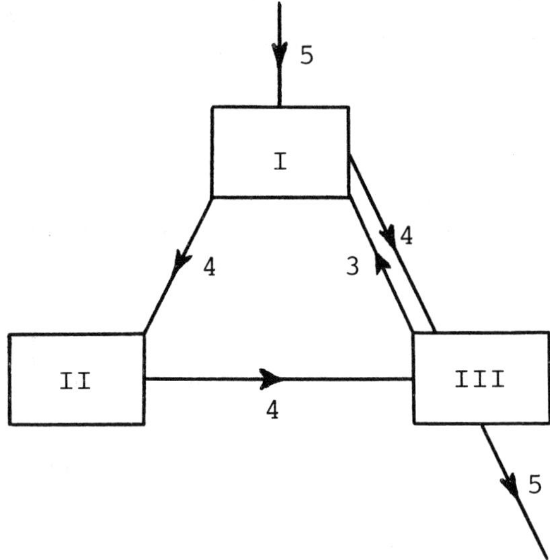

Initially, tank I contains 180 lb of salt while tanks II and III contain only pure water.

Let Y be the column 3-vector that gives the amount of salt in the tanks at time t.

a) [175] Derive the 3 by 3 matrix A so that $\frac{dY}{dt} = AY$.
Compare your answer with matrix 175.

b) Find the amount of salt in each tank at any time t.

c) Repeat parts a) and b) if the flow rates are changed to those in the figure below. In what way is the answer here different from the previous answer?

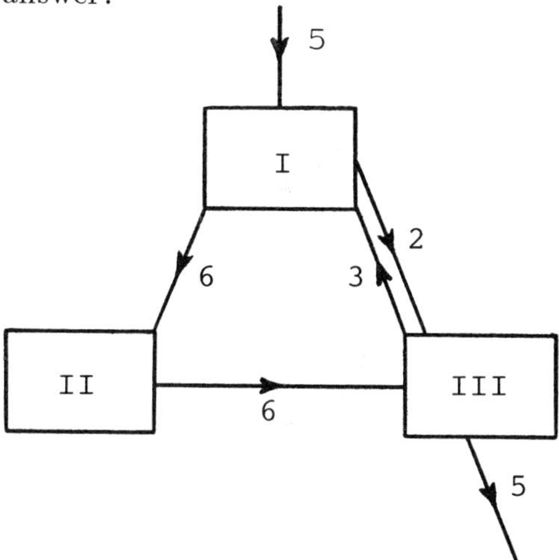

2. [176] Four tanks, each holding 120 gallons of liquid, are interconnected as in the figure. The numbers represent flow rates (in gal/min) with arrows indicating the direction of the flow. Pure water enters the system at tank I. Initially, tanks I and IV contain pure water, while tanks II and III each contain 15 grams of a toxic substance. Find the amount of toxic substance in each tank at any time t.

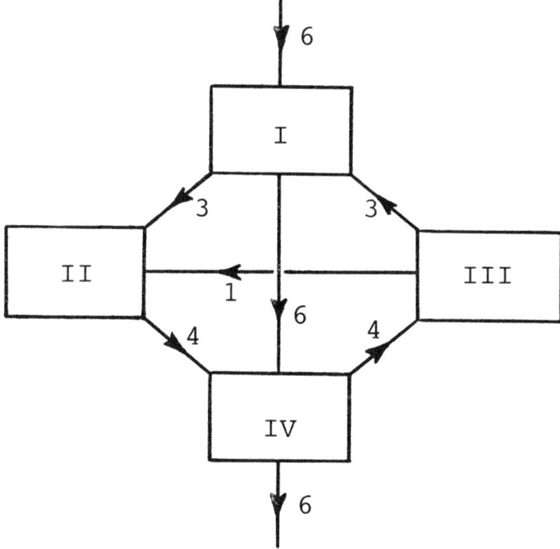

3. In a two-part mass-spring system as shown in the figure, choose the positive direction downward and let z_1 (resp. z_2) measure the displacement of weight m_1 (resp. m_2) from its equilibrium position.

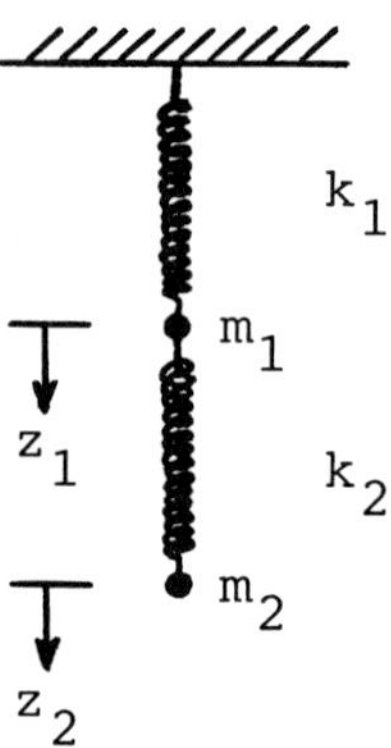

From Hooke's Law, z_1 and z_2 satisfy the system of second-order linear differential equations

$$m_1 \frac{d^2 z_1}{dt^2} = -k_1 z_1 + k_2(z_2 - z_1)$$

$$m_2 \frac{d^2 z_2}{dt^2} = -k_2(z_2 - z_1).$$

If we let $y_1 = z_1$, $y_2 = \frac{dz_1}{dt}$, $y_3 = z_2$, $y_4 = \frac{dz_2}{dt}$, we get the system of first-order linear differential equations

$$\frac{dy_1}{dt} = y_2$$

$$\frac{dy_2}{dt} = -\frac{k_1 + k_2}{m_1} y_1 + \frac{k_2}{m_1} y_3$$

$$\frac{dy_3}{dt} = y_4$$

$$\frac{dy_4}{dt} = \frac{k_2}{m_1} y_1 - \frac{k_2}{m_2} y_3.$$

By varying the k's and the m's, we can see the nature of the motion for different systems.

a) [177] Let $m_1 = m_2, k_1 = 6m_1, k_2 = 4m_2$. Solve the system subject to the initial conditions: at $t = 0, y_1 = y_2 = y_3 = 0, y_4 = 1$.

b) Let $m_1 = m_2, k_1 = 8m_1, k_2 = 3m_2$. Solve the system subject to the initial conditions: at $t = 0, y_1 = y_2 = y_3 = 0, y_4 = 1$.

c) What are the similarities and differences between the solutions in parts a) and b)?

4. Consider the mass-spring system shown in the figure.

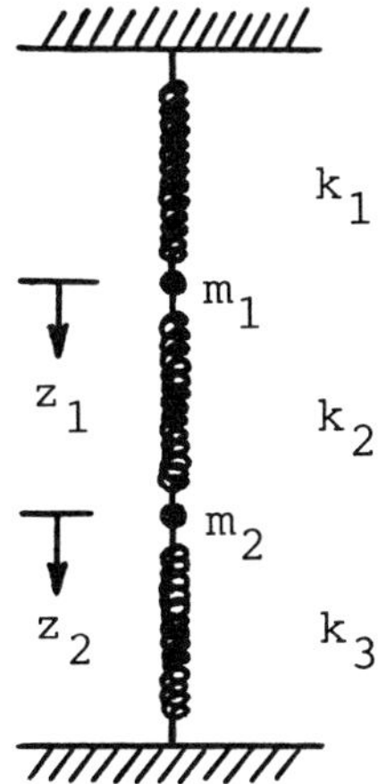

a) Using the notation in Problem 3, derive the system of four first-order differential equations that governs the motion of the weights.

b) [41] Let $m_1 = m_2, k_1 = 4m_1, k_2 = 3m_2, k_3 = k_1$. Solve the system subject to the initial conditions: at $t = 0, y_1 = y_2 = y_3 = 0, y_4 = 1$.

5. To study the distribution, breakdown, and synthesis of albumin in an animal, a researcher injects radioactive albumin into a vascular system where it mixes with the vascular fluid.

Let y_1, y_2, y_3, y_4 denote fractions of the amount originally injected that are present at time t in the vascular system (1), extravascular system (2), breakdown products (3), and excreted products (4), respectively.

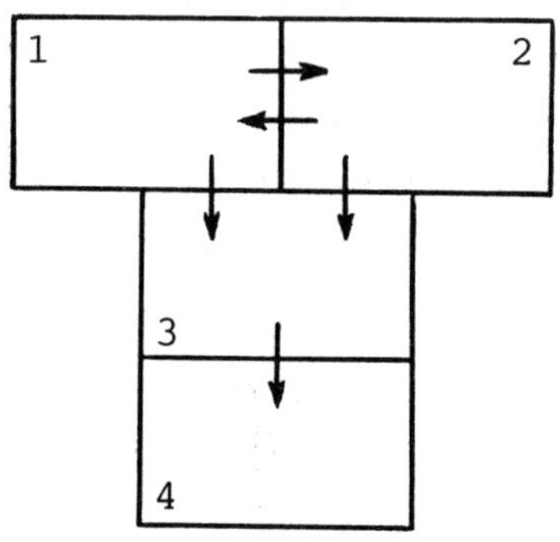

At $t = 0$, $y_1 = 1$, $y_2 = y_3 = y_4 = 0$. As $t \to \infty$, $y_1 = y_2 = y_3 = 0, y_4 = 1$. The rates of change of the y_i are

$$\frac{dy_1}{dt} = k_2 y_2 - (k_1 + k_3)y_1$$

$$\frac{dy_2}{dt} = k_1 y_1 - (k_2 + k_4)y_2$$

$$\frac{dy_3}{dt} = k_3 y_1 + k_4 y_2 - k_5 y_3$$

$$\frac{dy_4}{dt} = k_5 y_3$$

where the k's are rate constants that are determined experimentally and by the terminal conditions. By varying the k's, we can observe the flow of the radioactive substance under different conditions.

a) [178] Solve the system where $k_1 = k_2 = k_3 = k_4 = 1, k_5 = 5$.

b) Solve the system where $k_1 = 4, k_2 = k_3 = k_4 = 1, k_5 = 5$.

c) Solve the system where $k_1 = 4, k_2 = k_3 = 1, k_4 = 7, k_5 = 5$.

6. A compartment model is used to simulate the transport of a chemical compound (drug) through biological tissue. Consider a model with four compartments as shown in the figure.

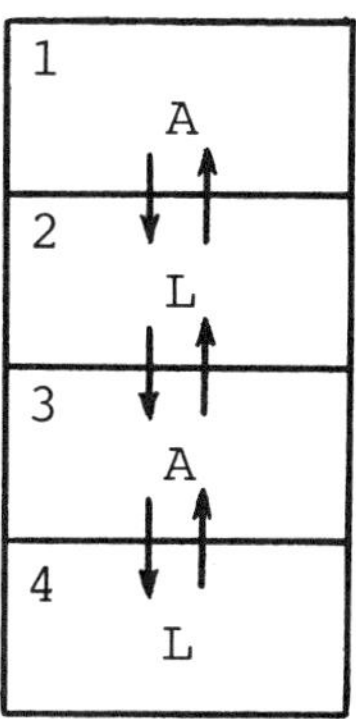

Let y_i be the drug concentration in compartment i at time t. Let a be the rate of transport from an aqueous compartment (A) to a lipid compartment (L) and let b be the rate from lipid to aqueous. If an initial quantity, say c_0, is injected in compartment 1, at $t = 0$ we have $y_1 = c_0, y_2 = y_3 = y_4 = 0$. (The units can be chosen so that $c_0 = 1$.)
The rates of change of the y_i are

$$\frac{dy_1}{dt} = -ay_1 + by_2$$

$$\frac{dy_2}{dt} = ay_1 - 2by_2 + ay_3$$

$$\frac{dy_3}{dt} = by_2 - 2ay_3 + by_4$$

$$\frac{dy_4}{dt} = ay_3 - by_4.$$

We can vary a and b and observe the flow of the drug concentration under different rates of transport.

a) [179] Solve the system where $c_0 = 1, a = .03, b = .04$. What are the proportions of the drug in the four compartments after a long period of time?

b) Solve the system where $c_0 = 1, a = .13, b = .0051$. What are the eventual proportions of the drug in the four compartments?

Appendix I

Descriptions of the Commands

In this appendix we give complete descriptions of all the commands. First, however, we explain how MAX searches through a full command to find the information it needs to carry out the described action. MAX looks at a full command as a sequence of words, separated by spaces, where a word is a sequence of characters that are not spaces. For example, command names and matrix names are words and thus cannot have imbedded spaces, while lists of parameter values are sequences of words and thus must be separated by spaces. MAX searches the full command from left to right for the command name, which must be the first word, and then for words that could be values for each of its parameters. The types of words it searches for are real numbers, non-negative integers, names of workspace matrices (with or without subscript expressions), file names (inside double quotes), and the special names I and ALL. A word that is none of these types and not a command name is a filler word, although there is an exception if this is the last word of the command. If MAX reaches such a last word while it is searching for a name to assign to the result matrix, it uses this word as the name. If, however, it does not find a word it can use as the name of the result, it assigns the default name.

NUMERICAL COMPUTATION

The numerical algorithms and concepts that underlie the ten commands in this first group are discussed in greater detail in Appendix II.

Rank — record the rank of Matrix1

computes and displays (but does not save) an upper-triangular matrix that has the same rank as the existing workspace matrix named Matrix1. MAX then asks you to enter the rank of Matrix1, which is recorded and used in the commands that require it: **solve, qrfactor, orthonormalize, nbasis,** and **cbasis**. Thus, one normally applies the **rank** command to a matrix before applying any of these other five commands. If Matrix1 is square, its determinant and r-condition (see the **determinant** command) are also displayed.

The rank of Matrix1 is simply the number of nonzero rows of the displayed upper-triangular matrix. However, round-off error can make it difficult to judge whether a row of this matrix is truly zero or not and hence make it difficult to judge the rank. The determinant and r-condition can help you make your judgment when Matrix1 is n by n: Its rank is less than n if and only if each of these numbers is zero. (Of course, these numbers are also affected by round-off error.)

Determinant — show the determinant and r-condition of Matrix1

computes and displays (but does not save) both the determinant and the r-condition (i.e., the reciprocal condition number) of the exisiting workspace matrix named Matrix1. Both numbers have the theoretical property that Matrix1 is singular if and only if this number is zero. The r-condition, however, is the more reliable measure, since it is independent of scaling and is less affected by round-off error.

Solve the system Matrix1 Matrix2 and store the solution in Matrix3

computes a solution of the linear system of equations $AX = K$, where A and K are the existing workspace matrices named Matrix1 and Matrix2, respectively, and stores the solution X in Matrix3. It uses the rank r of Matrix1 as follows. If A is r by r, then MAX uses the LU factorization of A to find the unique solution of the system; otherwise, it uses the QR factorization to find a least-squares solution (described in section 2.6). Since a least-squares solution need not be a solution, you may want to check your answer, which can be done either by checking if the rank of A equals the rank of the augmented matrix $\begin{pmatrix} A & K \end{pmatrix}$ or if $K - AX$ is zero. The value of r is whatever value you entered for the rank of Matrix1. If you have not yet entered the rank, this command will request it. **WARNING:** If you enter the wrong rank for Matrix1, the solution X will usually be incorrect.

Invert Matrix1 and store the inverse in Matrix2

computes the inverse of the existing workspace matrix named Matrix1 and stores the inverse in Matrix2.

LUfactor Matrix1 and store the L and U factors in [Matrix2 Matrix3]

computes the L and U factors of the existing square workspace matrix named Matrix1 and stores these factors in Matrix2 and Matrix3, respectively. MAX obtains the upper-triangular U factor by applying Gaussian elimination with partial pivoting to Matrix1. Therefore, the product of the lower-triangular factor L and the upper-triangular factor U is usually not Matrix1 but rather the result of having interchanged some rows of Matrix1 during the pivoting.

QRfactor Matrix1 and store the Q and R factors in [Matrix2 Matrix3]

computes the Q and R factors, not necessarily of the entire existing workspace matrix named Matrix1, but of the matrix consisting of r linearly independent columns of Matrix1 (where r is the rank of Matrix1) and stores the Q and R factors in Matrix2 and Matrix3, respectively. The columns of the Q factor are r orthonormal vectors in the column space of Matrix1, the R factor is an r by r upper-triangular nonsingular matrix, and their product is the matrix consisting of the r linearly independent columns of Matrix1. The value of r is whatever you entered for the rank of Matrix1. If you have not yet entered the rank, this command will request it. **WARNING:** If you enter the wrong rank for Matrix1 the columns of Matrix2 will not be an orthonormal basis of the column space of Matrix1.

Orthonormalize the columns of Matrix1 and store the result in Matrix2

computes r orthonormal vectors in the column space of the existing workspace matrix named Matrix1 (where r is the rank of Matrix1) and stores these vectors as the columns of Matrix2. The value of r is whatever value you entered for the rank of Matrix1. If you have not yet entered the rank, this command will request it. **WARNING:** If you enter the wrong rank for Matrix1, the columns of Matrix2 will not be an orthonormal basis of the column space of Matrix1.

NBasis — find a nullspace basis of Matrix1 and store in Matrix2

computes $n - r$ linearly independent vectors that lie in the nullspace of the existing workspace matrix named Matrix1 (where n is the number of columns and r is the rank of Matrix1) and stores these vectors as the columns of Matrix2. The value of r is whatever value you entered for the rank of Matrix1. If you have not yet entered the rank, this command will request it. **WARNING:** If you enter the wrong rank for Matrix1, the columns of Matrix2 will not be a nullspace basis of Matrix1.

CBasis — find a column space basis of Matrix1 and store in Matrix2

selects r linearly independent columns from among the columns of the existing workspace matrix named Matrix1 (where r is the rank of Matrix1) and stores these columns as the columns of Matrix2. The value of r is whatever value you entered for the rank of Matrix1. If you have not yet entered the rank, this command will request it. **WARNING:** If you enter the wrong rank for Matrix1, the columns of Matrix2 will not be a column space basis of Matrix1.

Eigenvalues and eigenvectors of Matrix1 — store in [Matrix2 Matrix3]

computes n eigenvalues and n corresponding eigenvectors for the existing n by n workspace matrix named Matrix1 and stores the eigenvalues as the first row of Matrix2 and the eigenvectors as the columns of Matrix3, if the eigenvalues are real. The eigenvalue in the jth column of Matrix2 corresponds to the eigenvector in the jth column of Matrix3. Eigenvectors are scaled so that the component with largest absolute value is 1.

When an eigenvector and its corresponding eigenvalue are complex, they are stored in a more complicated fashion, since MAX works only with real matrices. If the jth eigenvalue is complex with positive imaginary part, its real and imaginary parts are stored in the jth column and the first and second rows of Matrix2, and the real and imaginary parts of its corresponding eigenvector are stored in the jth and (j+1)st columns of Matrix3. The conjugate of this eigenvalue is stored in the (j+1)st column of Matrix2 but is not displayed, and its corresponding eigenvector is neither stored nor displayed.

Although Matrix2 always has two rows, a submatrix of the form Matrix2(j) will be interpreted as Matrix2(1, j) so that real eigenvalues are easy to refer to.

Unfortunately, the results of the **eigenvalues** command can be difficult to interpret when Matrix1 has fewer than n linearly independent eigenvectors. For then some of the n eigenvectors will be linearly dependent, but it might not be clear which. One check you can make is to find the rank and a nullspace basis for $xI - A$, where A is Matrix1 and x is this eigenvalue. However, round-off error may be so severe that you are unable to judge the value of x with sufficient accuracy.

ALGEBRA

Add Matrix1 to Matrix2 and store the sum in Matrix3

adds the existing workspace matrices named Matrix1 and Matrix2 and stores the sum in Matrix3.

Subtract from Matrix1 Matrix2 and store the difference in Matrix3

subtracts the existing workspace matrix named Matrix2 from the existing workspace matrix named Matrix1 and stores the difference in Matrix3.

Multiply Matrix1 (or Scalar1) by Matrix2 (or Scalar2) and store in Matrix3

multiplies the existing workspace matrices named Matrix1 and Matrix2, in that order, and stores their product in Matrix3. However, if a scalar is entered for Matrix1 or Matrix2, the product of this scalar and the other matrix is stored. A scalar can be a real number or a fraction or a 1 by 1 matrix.

Transpose Matrix1 and store the transpose in Matrix2

takes the transpose of the existing workspace matrix named Matrix1 and stores the transpose in Matrix2.

Compute Expression1 and store in Matrix1

computes the algebraic expression denoted Expression1 and stores the result in Matrix1. The expression may be constructed from the operations $+$, $-$, $*$, $/$, $'$, $\wedge$ and from real numbers, integers, and existing workspace matrices. Here $'$ denotes the transpose operation and $\wedge$ the exponentiation operation; exponents can be any integers. The order of precedence for the operations is: (1) $'$; (2) $\wedge$; (3) unary $+, -$; (4) $*, /$; (5) binary $+, -$. The letter I may be used in the expression to denote an identity matrix; MAX will determine its size from the context.

The algebraic expression in this command is parsed differently from the words and phrases in other commands: Spaces are permitted but not required within the expression, and the expression terminates when the next nonspace character or sequence of characters cannot be interpreted as part of a legitimate expression. Of course, the expression must be surrounded by spaces.

MODIFICATION

Edit Matrix1

permits you to change the entries of the existing workspace matrix named Matrix1 and simultaneously see this displayed matrix being changed. You change an entry by entering a new number in its place, where the number you type can be a real number in fixed or floating point notation or an integer or a quotient of two such numbers.

You can change only the entry at the *current position*, which is the position that the prompt at the bottom of the screen refers to and that is highlighted in the display. The initial current position is the upper left corner of the matrix. When you press the RETURN key, you get the next current position, which is the next lower one in the current column if you are not at the bottom of the column and is the top of the next column to the right if you are. If you are currently at the lower right corner of the matrix, you stay there. To leave the editor you give the subcommand **exit**. You can also move to a new current position via the arrow keys or other subcommands.

The complete set of subcommands is as follows:

Exit   leaves the editor and makes all the changes.
Cancel   leaves the editor but makes none of the changes.
Restore   restores all entries to their values before any of the changes.
Help   lists these subcommands.
Move n   moves the current position n units in the direction of the arrow
              (but stopping if the edge of the matrix is reached). RETURN
              is equivalent to down arrow (or right arrow in *rows* state).

Edge	moves the current position to the edge of the matrix in the direction of the arrow. RETURN is equivalent to down arrow (or right arrow in *rows* state).
Home	moves the current position to the upper left corner of the matrix (or the lower right corner if followed by the down or right arrow).
Rows	puts the editor into *rows* state, so that pressing the RETURN key changes the current position to the next one in that row (or the beginning of the next row).
Columns	puts the editor into *columns* state (the default state), so that pressing the RETURN key changes the current position to the next one in that column (or the beginning of the next column).
Decimal d	displays all real numbers with d digits after the decimal point. The default value of d is 2.

Change the entries in Matrix1

permits you to change the entries of the existing workspace matrix named Matrix1. It differs from the **edit** command in the following ways. The matrix is not displayed; the prompts are scrolled up the screen; the arrow keys do not function; the movement subcommands are therefore more limited; and there are two additional movement subcommands, **ahead n** and **back n**, with 1 as the default value of n in both of them.

Ahead n	sends the current position ahead n positions (but stopping if the lower right corner of the matrix is reached). Positions are counted in the order they are normally prompted for, assuming only the RETURN key is used to change the position.
Back n	sends the current position back n positions (but stopping if the upper left corner of the matrix is reached). Positions are counted in the order they are normally prompted for, assuming only the RETURN key is used to change the position.

Copy Matrix1 into Matrix2

copies the existing workspace matrix named Matrix1 into Matrix2.

Rows — copy rows IntegerList of Matrix1 into Matrix2

copies certain rows of the existing workspace matrix named Matrix1 into Matrix2. IntegerList is a list of positive integers that specify which rows of Matrix1 are to be copied and in what order.

Columns — copy columns IntegerList of Matrix1 into Matrix2

copies certain columns of the existing workspace matrix named Matrix1 into Matrix2. IntegerList is a list of positive integers that specify which columns of Matrix1 are to be copied and in what order.

Swap Matrix1 and Matrix2

swaps the corresponding entries of the existing workspace matrices named Matrix1 and Matrix2.

Augment Matrix1 by Matrix2 and store the result in Matrix3

forms the augmented matrix $\begin{bmatrix} \text{Matrix1} & \text{Matrix2} \end{bmatrix}$, where Matrix1 and Matrix2 are existing workspace matrices, and stores this augmented matrix in Matrix3.

Normalize the columns of Matrix1 and store the result in Matrix2

divides each column of the existing workspace matrix named Matrix1 by its Euclidean length (unless this length is zero) and stores the result in Matrix2.

Scale Matrix1 as a probability matrix and store the result in Matrix2

divides each column of the existing workspace matrix named Matrix1 by the sum of its entries (unless this sum is zero) and stores the result in Matrix2.

Round the entries of Matrix1 and store the result in Matrix2

rounds each entry of the existing workspace matrix named Matrix1 to the nearest integer and stores the result in Matrix2.

Rename Matrix1 as Matrix2

changes the name of the existing workspace matrix named Matrix1 to Matrix2.

Delete Matrix1 (or list of matrices)

deletes from the workspace the existing workspace matrix named Matrix1 (or a list of existing workspace matrices). Delete ALL causes all the existing workspace matrices to be deleted.

Clear the workspace of all matrices

is equivalent to Delete ALL.

CONSTRUCTION

Identity — store the size Integer1 identity matrix in Matrix1

constructs an identity matix with Integer1 rows and columns and stores it in Matrix1.

Zero — store the Integer1 by Integer2 zero matrix in Matrix1

constructs a zero matrix with Integer1 rows and Integer2 columns and stores it in Matrix1.

Ones — store the Integer1 by Integer2 matrix of ones in Matrix1

constructs a matrix of ones with Integer1 rows and Integer2 columns and stores it in Matrix1.

Diagonal — store Matrix1 diagonally in Matrix2

constructs a square matrix with zeros as its off-diagonal entries and with its diagonal entries copied from the single column or single row of the existing workspace matrix named Matrix1. This diagonal matrix is stored in Matrix2.

Create a random Integer1 by Integer2 matrix of rank Integer3 — store in Matrix1

creates a matrix with Integer1 rows and Integer2 columns of rank Integer3 with random integer entries, and stores it in Matrix1.

INFORMATION

Help

displays a list of all the commands and then asks which command you want help on. The help is simply the description of the command given in this appendix, though occasionally abbreviated. You can skip over the list of the commands by entering **help** followed by the name of the command you want help on. Also, you can request help on any of the underlined terms discussed in the Overview or this appendix; in place of a command name, enter at least the first three letters of the term, preceded by a / with no spaces between. In this case, the help consists of the relevant paragraph in the Overview or this appendix.

Display Matrix1 (or list of matrices)

displays on the screen the existing workspace matrix named Matrix1 (or a list of existing workspace matrices). If the matrix is too large for the screen, it is displayed as if it were partitioned into submatrices of a size that does fit on the screen, with the first row of the partition displayed first, from left to right, then the second row, and so on. The number of decimal places displayed on the screen is governed by the **decimal** command and does not necessarily reflect the number of places to which numbers are kept in the workspace.

Complex eigenvalues and eigenvectors are not displayed in the manner they are stored; their real and imaginary parts are stored as real numbers, but they are displayed as complex numbers.

Print Matrix1 (or list of matrices) to "File1"

prints the existing workspace matrix named Matrix1 (or a list of existing workspace matrices) in the file named File1. These matrices appear in the file in exactly the form they would if you had used the **display** command instead. The default extension for the file name is .TXT. If you are printing wide matrices and your printer is more than 80 columns wide, you should first use the **width** command to set MAX's width to the width of your printer. For tall matrices, increase the **height**. When you quit MAX, you can send File1, which is simply an ASCII text file, to a printer by using any print facility on your system. On many systems, you can print File1 on your attached printer without leaving MAX, by using a special filename such as "PRN" in the print command.

List

lists the names of all the existing workspace matrices, gives the number of rows and columns of each, and gives the rank of each one whose rank has been recorded.

Workspace

reports the amount of room still available in the workspace. However, you may be told that there is not enough room left in the workspace for a command, even when the number of entries of the result matrix is within the amount of room reported by the **workspace** command, since many commands require temporary use of additional workspace while they are being carried out.

FILES

Get matrix Integer1 and store it in Matrix1

gets the matrix numbered Integer1 from group storage and stores it in Matrix1. The default name of Matrix1 is Mk, where k is Integer1.

Put Matrix1 (or list of matrices) in (or starting in) storage location Integer1

puts the existing workspace matrix named Matrix1 into group storage in storage location Integer1, or puts a list of existing workspace matrices into consecutive storage locations starting in location Integer1. It will overwrite any matrix that occupies one of these group storage locations.

Group storage is a collection of ten files, named GRP00.DAT through GRP09.DAT, each of which may contain up to twenty matrices. The matrices in GRP00.DAT are numbered 1 through 20, those in GRP01.DAT are 21 through 40, and so on. If there are fewer than 20 matrices in one of the ten files, they are assigned the lowest of the 20 associated numbers, and the higher numbers are not assigned to any matrices. Thus, there can be gaps in the numbering of group storage matrices, but only at the ends of the ten files.

Since the group storage files are ASCII text files, you may edit them. However, you must retain their structure: Each matrix is preceded by two integers that denote its number of rows and columns, respectively; the first row of a matrix precedes its second row, etc.; and each file ends with two integer zeros.

Save Matrix1 as "File1"

saves a copy of the existing workspace matrix named Matrix1 in the file named File1. The default extension for the file name is .MAT. Since the result is an ASCII text file, you may edit it. However, if you want to be able to **retrieve** this file, you must retain its structure as follows. The file contains exactly one matrix, and its entries are preceded by two integers that denote its number of rows and columns, respectively. The first row of the matrix precedes the second row, and so on.

Retrieve "File1" and store it in Matrix1

retrieves the matrix from the existing file named File1 and stores it in Matrix1. The default extension for the file name is .MAT. The default name of Matrix1 is the file name, without its extension. The structure of the file must be as described for the **save** command.

Store the workspace as "File1"

stores a copy of the entire workspace in the file named File1. The default extension for the file name is .WOR. The resulting file is not designed to be edited.

Restore the workspace from "File1"

uses the file named File1 to restore the workspace to the state it was in when it was stored in this file. Any matrices that are already in the workspace will be overwritten by this command. The default extension for the file name is .WOR.

Script Files

The next four commands concern <u>script files</u>, which are files that can be used either for classroom demonstrations or as a way to execute a frequently used sequence of commands with a single command. A script file consists of a list of responses to prompts for commands and command parameters, one response per line. It must not contain responses to other kinds of prompts, not even the prompt that asks for the rank of a matrix.

A script file may also have script parameters, which are something like procedure parameters in a programming language. If a script has such parameters, they must be listed in the top line of the file, separated by spaces. A script file with no parameters must have its top line blank. To add parameters to a script file, you must use an editor. (Script files are ASCII text files whose default extension is .MAC.)

When you **demonstrate** or **execute** a script file, you are first prompted for values of all the script parameters. MAX then replaces each occurence of these script parameters in your script file by the value you supplied. The replacement is accomplished by a string substitution, which is carried out even if the parameter occurs in the middle of a word. Thus, to prevent the parameter values from being inserted in more places than you intended, it is wise to begin and end your script parameter names with unusual characters, as in the second of the following two examples. The first example is a script file that was created within MAX via the **open** and **close** commands. The second is an edited version that includes script parameters. Both script files do a linear least-squares fit to a set of points (x, y), but the first is specific and the second is more general.

The script file "LS"

```
get 152 data
ones 8 1
aug % data data
noquery
sol data(,1-2) data(,3) soln
print data(,2-3) soln "data"
```

The script file "LSparam"

```
$gsNum$ $rows$ $DataName$ $RsltName$ $FlName$
nowarn
noshow
noecho
get $gsNum$ $DataName$
ones $rows$ 1
aug % $DataName$ $DataName$
noquery
sol $DataName$(,1-2) $DataName$(,3) $RsltName$
print $DataName$(,2-3) $RsltName$ "$FlName$"
warn
show
echo
query
```

Demonstrate the script in "File1"

executes all the commands in the script file named File1, but it displays each response in the file and waits for you to press the RETURN key before it executes the command. If you type **cancel** before pressing the RETURN key, the rest of the demonstration is canceled. If the initial line of File1 is not a blank line, MAX first prompts you for values of the parameters listed in that line. For more information, see the above description of script files.

Execute the script in "File1"

executes all the commands in the script file named File1. If MAX detects an error in a command, the rest of the execution is canceled. If the initial line of File1 is not a blank line, MAX first prompts you for values of the parameters listed in that line. Since **execute** is normally used to carry out a sequence of commands as quickly as possible, you will normally want to cut down on the displays and the prompts. Thus, you may want to first give the **noshow, nowarn,** and **noecho** commands. You may even want to give the **noquery** command, if you know that all the matrices whose rank is needed either have maximal rank or will have had their rank recorded in advance. For more information, see the above description of script files.

Open "File1" for a script

creates an empty file named File1 and then inserts a blank line into the beginning of the file. Thereafter, until you give the **close** command, all your responses to the command prompts and parameter prompts are stored in File1. Responses to other prompts are not stored in the file. For more information, see the above description of script files.

Close the currently open script file

closes the currently open script file, if there is one.

CHANGE OF STATE

Bell

returns MAX to the default *bell* state. Thereafter, whenever MAX detects an error in a command that you give, a warning bell sounds.

NoBell

takes MAX out of the default *bell* state. Thereafter, no warning bell sounds when MAX detects an error in a command that you give.

Show

returns MAX to the default *show* state. Thereafter, all result matrices are automatically displayed on the screen, unless your command ends in the special character #.

NoShow

takes MAX out of the default *show* state. Thereafter, no matrices are displayed on the screen, except for matrices you specifically ask to be displayed by the **display** command. Also, the messages normally displayed by the **put** command and the change of state commands are suppressed.

Echo

returns MAX to the default *echo* state. Thereafter, all commands you give are echoed on the screen.

NoEcho

takes MAX out of the default *echo* state. Thereafter, no commands you give are echoed on the screen.

Warn

returns MAX to the default *warn* state. Thereafter, MAX gives a warning whenever you are about to lose all your workspace matrices or are about to overwrite one of them. It also gives you the opportunity to cancel the command that is about to cause this to happen.

NoWarn

takes MAX out of the default *warn* state. Thereafter, no warning is given when you are about to lose all your workspace matrices or are about to overwrite one of them.

Query

returns MAX to the default *query* state. Thereafter, you will be queried for the rank of Matrix1 in the **solve, qrfactor, orthonormalize, nbasis,** and **cbasis** commands, unless you entered the rank of this matrix in an earlier command.

NoQuery

takes MAX out of the default *query* state. Thereafter, you will not be queried for the rank of Matrix1 in the **solve, qrfactor, orthonormalize, nbasis,** and **cbasis** commands. Instead, MAX will assume that this matrix has the maximal rank for a matrix of that size, unless you entered the rank of the matrix in an earlier command.

Exit

> takes you out of MAX. All workspace matrices are lost.

Cancel

> cancels a command before it can be carried out. The appropriate time to give the **cancel** command is in response to any prompt. Also, a command can be canceled by simply pressing the RETURN key in response to a prompt for a parameter that has no default value.

Decimal — set decimal places to Integer1

> causes all subsequently displayed real numbers to be printed with Integer1 digits after the decimal point. The default value of Integer1 is 2. All numbers whose absolute value is greater than or equal to 0.1 and less than or equal to 9999 are written in fixed point notation; all others are written in floating point. However, if Integer1 is larger than 4, all numbers are written in floating point.

Height — set height to Integer1 lines

> causes all subsequently displayed matrices to fit within 4 fewer lines than Integer1, since 4 lines are reserved for other information. In the **edit** command, the displayed matrix fits within 8 fewer lines than Integer1. The default value of Integer1 is 24.

Width — set width to Integer1 columns

> causes all subsequently displayed matrices to fit within Integer1 columns. In the **edit** command, the displayed matrix fits within 5 fewer columns than Integer1. The default value of Integer1 is 80.

Upper Bounds on the Variables Used in MAX

4096 matrix entries in the workspace.
  99 rows or columns in a workspace matrix.
  80 characters in a line of input from the keyboard or a script file; excess characters are ignored.
  63 characters in a file name, including the path name.
  10 characters in a matrix name.
  10 characters in a script parameter name.
   8 parameters in a script file.
  25 characters in a script parameter value.
  19 parameters in a command.
   4 digits in an integer value for a parameter.
1E38 for the magnitude of a real number.

Synonyms for Selected Commands

directory	for **list**
edt	for **edit**
logout	for **exit**
name	for **rename**
quit	for **exit**

Appendix II

The Numerical Algorithms in MAX

The ten MAX commands that perform numerical computations are based on algorithms that have been used extensively in modern scientific computing, notably in the LINPACK and EISPACK subroutine packages [Dongarra et al (1979), Smith et al (1976)]. Many of the algorithms appeared earlier in book form [Wilkinson and Reinsch (1971)] and earlier yet in journal articles, with improvements added at each stage. We will discuss here only the few of these algorithms used in MAX, and our discussion will provide only an outline of how they work. Much more information is available in the above references and in textbooks such as [Stewart (1973), Strang (1980)].

The LINPACK algorithms for calculating and using the LU factors of a square matrix are the basis for the **invert, determinant,** and **LUfactor** commands, and for the **solve** command when the coefficient matrix is nonsingular. The LINPACK algorithms for calculating and using the QR factors of an arbitrary matrix are the basis for the **rank, orthonormalize, QRfactor, nbasis,** and **cbasis** commands, and for the **solve** command when the coefficient matrix is singular. The **eigenvalues** command is based on the EISPACK algorithms for finding the complete set of eigenvalues and eigenvectors of an arbitrary real square matrix.

The LU factorization of a square matrix A is accomplished as follows. The matrix A is reduced to an upper-triangular matrix U by Gaussian elimination with partial pivoting. That is, before reducing the jth column of A, a row interchange is performed so that the diagonal entry (the pivot) is at least as large in absolute value as all the entries below it. Then the pivot is used to introduce zeros below the diagonal by multiplying the jth row by $-a_{ij}/a_{jj}$ and adding it to the ith row, for each i greater than j. (If $a_{jj} = 0$, this step is omitted.) The lower-triangular matrix L is formed by putting 1's on the diagonal and the numbers a_{ij}/a_{jj} below the diagonal (but permuted so they appear in their original positions before the row interchanges.) The product of L and U is A with some of its rows permuted, where the permutation is determined by the row interchanges that occured during the reduction. Thus, $PA = LU$ for some permutation matrix P.

If A is nonsingular, the system $AX = K$ is solved by treating A as the product $P^{-1}LU$ and then solving $P^{-1}LUX = K$ in two stages. First $(P^{-1}L)C = K$ is solved for C by applying the same row operations to K as were earlier applied to A; that is the meaning of $C = (P^{-1}L)^{-1}K$. Then $UX = C$ is solved for X by back-substitution. This method of solving $AX = K$ is quite similar to the hand method that employs the augmented matrix.

The inverse of A is computed as the product $U^{-1}(P^{-1}L)^{-1}$, where U^{-1} and $(P^{-1}L)^{-1}$ are found as follows. The inverse $(P^{-1}L)^{-1}$ is found by applying the same operations to the identity matrix as were earlier applied to A, and U^{-1} can be found by a process akin to back-substitution, since it is the solution of $UX = I$. Specifically, U, U^{-1} and I are partitioned into

$$U = \begin{pmatrix} U_1 & U_2 \\ 0 & u_3 \end{pmatrix}, \quad U^{-1} = \begin{pmatrix} V_1 & V_2 \\ 0 & v_3 \end{pmatrix}, \quad I = \begin{pmatrix} I_1 & 0 \\ 0 & 1 \end{pmatrix}$$

where the lower right submatrices are all 1 by 1. Then from $UU^{-1} = I$, it follows that

$$v_3 = \frac{1}{u_3}, \quad V_1 = U_1^{-1}, \quad V_2 = -v_3 V_1 U_2.$$

The algorithm continues by computing U_1^{-1} in this same way.

The determinant of A is simply the product of the diagonal entries of U, multiplied by -1 if there were an odd number of row interchanges during the reduction.

The last of the LINPACK algorithms relating to LU factorization that we discuss is the algorithm for finding the reciprocal condition number, which is referred to in MAX as the r-condition. The *condition number* of a nonsingular matrix A is defined to be

$$\kappa = \| A \| \| A^{-1} \|$$

where the matrix norm $\| A \|$ is defined by

$$\max \frac{\| AX \|}{\| X \|}$$

and the maximum is taken over all nonzero column vectors X. The vector norm $\| X \|$ is usually the Euclidean length, but in LINPACK it is

$$\| X \| = \sum_j | x_j |$$

since this is faster to compute and since the matrix norm is then simply

$$\| A \| = \max_j \| A_j \|$$

where A_j denotes the jth column of A. The computation of $\| A^{-1} \|$ is not so simple. In fact, we do not even want to assume that A^{-1} exists, since we want a measure that will help us distinguish singular matrices from nonsingular ones. Note, however, that

$$\frac{1}{\| A^{-1} \|} = \frac{1}{\max \| A^{-1}X \| / \| X \|} = \min \frac{\| AY \|}{\| Y \|}$$

and that this last expression is defined even when when A is singular. We therefore define the *r-condition* of any square matrix A to be

$$\frac{1}{\|A\|} \, \min \, \frac{\|AY\|}{\|Y\|}$$

and we note that it equals $1/\kappa$ when A is nonsingular. The above minimum may be approximated by choosing X so that $\|X\| / \|Y\|$ is as small as possible, where Y is a solution of $AY = X$. How this is done we leave to the references.

The QR factorization of a matrix A is accomplished, as in the LU factorization, by reducing A to an upper-triangular matrix. The reduction process, however, is not done by elementary row operations but by orthogonal transformations. Specifically, A is multiplied on the left by matrices of the form

$$H = I - 2UU', \quad \text{where} \quad \|U\| = 1,$$

which are orthogonal matrices called *Householder transformations* or *elementary reflectors*. For example, to introduce all zeros below the $(1,1)$ entry of A, we multiply A on the left by

$$H_1 = I - 2U_1 U_1',$$

where

$$U_1 = k\left(E_1 + \frac{A_1}{\sigma \|A_1\|}\right)$$

and where E_1 is the first column of the identity matrix, σ is ± 1 according as the $(1,1)$ entry of A is positive or negative, and k is the positive scalar that forces $\|U_1\|$ to be 1. To determine the Householder transformation H_2 that is multiplied on the left of $H_1 A$, we ignore the first row and column of $H_1 A$ and proceed as before. Thus, A can be written as the product:

$$A = (H_1^{-1} H_2^{-1} \ldots)R = QR$$

where R is the upper-triangular matrix that A is reduced to and Q is the product of the inverses of the Householder transformations.

If A is m by n and has linearly independent columns, this factorization can be written in a more useful form:

$$A = QR = \begin{pmatrix} Q_1 & Q_2 \end{pmatrix} \begin{pmatrix} R_1 \\ 0 \end{pmatrix} = Q_1 R_1$$

where Q_1 is m by n and has orthonormal columns and where R_1 is n by n, nonsingular and upper-triangular. This is the factorization produced by the **QRfactor** command, and Q_1 is the result produced by the **orthonormalize** command.

If A does not have linearly independent columns, it is useful to rearrange the columns of A during the reduction. Specifically, before H_1 is computed, the largest (in norm) column of A is interchanged with the first column; then, in the matrix obtained from $H_1 A$ by ignoring the first row and column, the largest column is again interchanged with the first; and so on. These column interchanges have the effect of moving the linearly independent columns of A to the left.

This variation of QR factorization is used in the **rank, nbasis,** and **cbasis** commands. Specifically, the matrix displayed by the **rank** command is the R factor of the QR factorization of A in which columns are so interchanged. Thus, if the columns of A, and hence R, are linearly dependent, the independent columns will be moved to the left, which makes it easier to determine the rank of R and hence of A. The **cbasis** command simply selects the leftmost of the rearranged columns of A, where the number of columns it selects is the rank of the matrix. The **nbasis** command produces a nullspace basis by expressing each linearly dependent column of A as a linear combination of the linearly independent columns. For example, if A has the columns A_1, A_2, A_3, A_4, A_5 and if A_1, A_3, A_5 are the linearly independent ones (as determined by the QR factorization process), the equations

$$A_2 = aA_1 + bA_3 + cA_5$$
$$A_4 = dA_1 + eA_3 + fA_5$$

can both be solved, and so $\begin{pmatrix} -a \\ 1 \\ -b \\ 0 \\ -c \end{pmatrix}$ and $\begin{pmatrix} -d \\ 0 \\ -e \\ 1 \\ -f \end{pmatrix}$ constitute a nullspace basis for A.

We have not yet explained how the above equations are solved. The method is exactly that used in the **solve** command when the coefficient matrix is singular, so we will describe it only once. Suppose we want to find the least-squares solution of $AX = K$, where A has linearly independent columns, and suppose we have the factorization $A = QR$, where Q has orthonormal columns and R is nonsingular and upper-triangular. Then, as described in Section 2.6, we must solve the normal equation

$$A'AX = A'K$$
$$\text{i.e.} \quad (QR)'(QR)X = (QR)'K$$
$$\text{i.e.} \quad (R'Q'QR)X = (R'Q'K$$
$$\text{i.e.} \quad R'RX = R'Q'K \qquad \text{since } Q'Q = I$$
$$\text{i.e.} \quad RX = Q'K \qquad \text{since } R \text{ is nonsingular.}$$

Then X can be found by back-substitution, since R is upper-triangular.

The EISPACK algorithms used in the **eigenvalues** command are especially intricate, and our descriptions of them will be especially cursory. Suppose A is a square matrix whose eigenvalues and eigenvectors are to be found. We will make frequent use of the fact that A has the same eigenvalues as any similar matrix, that is, any matrix of the form $P^{-1}AP$. For example, in the first step of the method, the matrix A is transformed to a similar matrix that is both balanced and upper-Hessenberg. (A *balanced* matrix is one in which each row has the same length as the corresponding column; an *upper-Hessenberg* matrix is one that has all zeros below the subdiagonal — the diagonal just below the main diagonal.)

This transformation is preliminary to the central part of the method and is done to speed up the central part and make it numerically more stable. This central part, which we describe now, is called the QR algorithm, since it makes repeated use of QR factorization.

The QR algorithm is iterative; it produces a sequence of matrices

$$A_0, A_1, A_2, \ldots$$

that are all similar to A and that converge to a limit whose eigenvalues are easy to find. A_0 is defined to be A, and A_{k+1} is defined for $k \geq 0$ by

$$A_{k+1} = R_k Q_k$$

where

$$A_k = Q_k R_k$$

is the QR factorization of A_k. Note that

$$A_{k+1} = Q_k'(Q_k R_k)Q_k = Q_k' A_k Q_k$$

which is similar to A. Also, it can be shown that the matrices A_k are upper-Hessenberg and, most importantly, that they converge to a *quasi-triangular* matrix. This is a matrix that is almost upper-triangular but can have 2 by 2 blocks, not just 1 by 1 blocks, along its diagonal. The 1 by 1 blocks are the real eigenvalues of A, and the eigenvalues of the 2 by 2 blocks are the complex eigenvalues of A. Moreover, the eigenvectors of this quasi-triangular matrix can be found by back-substitution and can be transformed into the eigenvectors of A by applying the similarity transformation that relates the two matrices.

Actually, the version of the QR algorithm in EISPACK is more complicated yet. It incorporates a shift that speeds up convergence further. Instead of factoring A_k, we factor a shifted matrix of the form $A_k - c_k I$, where c_k is a scalar that is determined from the elements of A_k and that converges toward an eigenvalue of A as k increases. Then A_{k+1} is defined by

$$A_{k+1} = R_k Q_k + c_k I$$

where

$$A_k - c_k I = Q_k R_k$$

is the QR factorization of $A_k - c_k I$. Note that

$$
\begin{aligned}
A_{k+1} &= Q_k' Q_k R_k Q_k + c_k I \\
&= Q_k'(A_k - c_k I)Q_k + c_k I \\
&= Q_k' A_k Q_k.
\end{aligned}
$$

This version of the QR algorithm produces a sequence in which one eigenvalue (or one pair of complex conjugate eigenvalues) is approached quite rapidly and the others more slowly. When one eigenvalue (or pair of conjugate eigenvalues) has been isolated, the algorithm then continues with the smaller matrix that is left after the row(s) and column(s) containing it are discarded. By acting on matrices of decreasing size, a considerable speed-up is achieved.

Appendix III
Group Storage

```
M1
      -5.          1.          5.
      -7.          4.          4.
      -1.          1.          1.

M2
       2.          2.          1.
       3.          0.          4.
       1.         -1.          2.

M3
       5.          0.          3.
       2.          1.         -7.
      -1.         -1.          1.
       1.         -1.          2.

M4                            M5
       1.          2.          3.          3.         -1.          2.
       3.          2.          1.          2.          1.          1.
       0.          2.          4.          1.          3.          0.

M6                            M7
       1.          1.          3.          1.         -2.          1.
      -1.          1.          0.          2.         -5.          0.
       2.          2.          0.         -1.          3.          1.
       0.          0.          1.          2.          0.         10.

M8
       3.          2.          2.
       1.          4.          1.
      -2.         -4.         -1.

M9
     -11.         -9.          6.
      20.         16.        -10.
       8.          6.         -3.
```

Page 120

M10
```
    -22.        -3.        -25.
     40.         6.         44.
     16.         3.         16.
```
M11
```
      3.         1.         -5.
      9.         3.          4.
     -2.         7.         -1.
```
M12
```
      1.         1.          1.          1.
      1.         3.          2.          4.
      2.         0.          1.          1.
```
M13
```
      2.        -1.         -1.          2.
     -1.         2.          1.         -1.
      1.         4.          1.          1.
```
M14
```
      1.        -1.          2.
      3.        -3.          6.
      1.        -1.          2.
```
M15
```
      1.         0.          0.         -1.
      1.         1.          0.         -4.
      0.         1.          1.         -6.
      0.         0.          1.         -3.
```
M16
```
      2.         1.          0.
      0.         2.          0.
      0.         0.         -4.
```
M17
```
      1.        -2.          1.
      3.        -5.          0.
     -2.         6.          7.
```
M18
```
      1.        -1.          1.
      1.         0.         -1.
      0.         1.          1.
```
M19
```
      1.         0.          1.          0.
      2.         1.          0.          0.
      0.         0.          1.          1.
      1.        -2.          0.         -5.
```

```
M20
      2.          -1.          -1.           2.
     -1.           2.           1.          -1.
      1.           3.           2.           3.
      2.          -3.          -1.           4.
M21
     36.           1.           1.
     31.           3.           1.
      1.           1.           1.
M22                                M23
      1.          -1.           0.           5.           0.           2.
      0.           0.           1.           0.           1.           0.
     -1.           2.           0.          -4.           0.          -1.
M24
      2.          -1.           1.
     -1.           2.          -1.
      1.          -1.           2.
M25                                M26
     17.          -8.         -12.          14.           3.
     46.         -22.         -35.          41.          -5.
     -2.           1.           4.          -4.          -4.
      4.          -2.          -2.           3.
M27                                M28
      1.           1.           1.           0.          14.
      2.          -2.           0.           2.         -41.
      4.           4.          -4.           0.          -1.
      8.          -8.           0.          -8.         -61.
M29          M30
      1.           4
      1.           0
      1.          -3
              5
M31                                M32
      2.           4.           3.           2.          -1.
      3.           6.           5.           2.         -11.
      2.           5.           2.          -3.         -10.
      4.           5.          14.          14.          40.
M33                                M34
      0.           1.           3.          -2.         -47.
      1.           2.           6.           0.         -35.
      2.           3.           9.           2.         -13.
      1.           1.           3.           2.          27.
```

M35   M36   M37   M38   M39
    1.   1.   5.   0.   0.   -9.   10.   7.
    1.   0.   -3.   6.   3.   0.   -4.   2.
    0.   1.   2.   0.   0.   -9.   8.   1.
    1.   0.   7.   -5.   -2.   1.   5.   10.

M40
    0.   3.   1.
    0.37   0.   0.
    0.   0.3   0.

M41
    0.   1.   0.   0.
   -7.   0.   3.   0.
    0.   0.   0.   1.
    3.   0.   -7.   0.

M42   M43   M44
    9.   3.   4.   0.   3.
    3.   1.   -5.   -19.   1.
   -2.   7.   -1.   84.   0.
                                                               1.

M45   M46   M47
    1.   3.   2.   1.   -1.   7.   -1.
    7.   -3.   2.   1.   -7.   1.   1.
    3.   1.   2.   1.   -3.   5.   -2.

M48
   -2.   5.   1.   1.
    1.   -3.   5.   -1.
   -1.   1.   7.   -1.
    1   -2.   -6.   0.

M49
   -5.   2.   -1.   1.   -9.
    7.   3.   13.   3.   17.
    1.   -5.   -9.   8.   10.
    3.   8.   19.   4.   10.

M50
   -7.   9.   4.   5.   3.
    5.   1.   0.   -2.   1.
    3.   11.   4.   1.   5.
   -5.   -1.   0.   2.   -1.

```
M51                  M52                  M53
    2.                   0.                  -7.
    2.                  -5.                   2.
    3.                  -1.                  -8.
   11.

M54                                                   M55
   -1.    5.    7.    1.      2.      4.
    3.   -2.    5.   10.      7.      1.
    7.    1.   23.   29.     22.      8.

M56                                                   M57
   -1.    3.   -5.    4.     18.      4.
    1.   -2.    4.    0.     -7.     -3.
    2.    0.    4.   -3.     -8.     -2.
    5.    1.    9.    2.      2.     -4.

M58
    1.    0.    1.    0.      4.
    2.    1.    0.    1.      0.
    0.    0.    1.   -1.     -3.
    1.   -2.    0.    3.      5.

M59
    1.    1.    2.   -1.
    2.    4.   -3.    1.
    3.   -1.   -5.    6.

M60
    1.    2.    3.    1.
    3.    2.    1.    1.
    0.    2.    4.    1.

M61
    1.    2.    3.    1.
    0.    1.    4.   -1.
    3.    0.  -15.    9.
    5.    8.    7.    7.

M62
    0.    0.    3.    4.
    1.    3.    1.   -5.
    6.   -2.    7.   -1.

M63                                      M64         M65
    1.    3.   -5.    0.        1.           3.
    3.    9.    4.    1.      -18.          -6.
   -6.   -2.   -1.    6.      -67.          -5.
   -1.    1.   -3.    2.      -17.
```

M66

3.	-2.	0.
3.	-1.	5.
-4.	-3.	-6.

M67

1.	3.	7.	10.
2.	0.	-1.	5.
3.	-2.	-1.	-3.
-4.	2.	1.	0.

M68

0.24	0.2	0.3
0.4	0.16	0.42
0.16	0.12	0.1

M69

70.
152.
50.

M70

1.	1.	-1.
6.	0.	5.
0.	3.	5.

M71

0.
4.
19.

M72

1.	1.	1.
0.	2.	1.
-2.	0.	-1.
0.	-3.	-2.

M73

0.
1.
1.
1.

M74

-1.	5.	3.	0.
3.	1.	7.	1.
0.	1.	1.	-2.
2.	-3.	1.	4.

M75

7.	1.	-17.
12.	1.	7.
0.	0.	-11.
4.	1.	29.

M76

1.	1.	0.
2.	0.	1.
0.	1.	0.
1.	0.	-2.

M77

0.	0.	4.	0.
1.	1.	0.	0.
-1.	-1.	-3.	1.
-3.	3.	-5.	-5.

M78

1.	-3.	0.	1.	9.	-11.
-1.	2.	-3.	-3.	-5.	9.
1.	1.	12.	9.	-7.	-3.
6.	-5.	1.	-6.	2.	-2.

M79

1.	-4.	3.	3.
7.	0.	6.	1.
2.	12.	-5.	-6.
1.	-1.	2.	-4.

M80

```
   1.          -3.          2.
  -1.           2.          3.
   1.           1.         -4.
```

M81   M82

```
0.      0.25   0.      0.25   0.      0.      0.      0.      0.      2.5
0.25    0.     0.25    0.      0.25    0.      0.      0.      0.      2.5
0.      0.25   0.      0.      0.      0.25    0.      0.      0.      2.5
0.25    0.     0.      0.      0.25    0.      0.25    0.      0.      0.
0.      0.25   0.      0.25    0.      0.25    0.      0.25    0.      0.
0.      0.     0.25    0.      0.25    0.      0.      0.      0.25    0.
0.      0.     0.      0.25    0.      0.      0.      0.25    0.      0.
0.      0.     0.      0.      0.25    0.      0.25    0.      0.25    0.
0.      0.     0.      0.      0.      0.25    0.      0.25    0.      0.
```

M83   M84

```
0.      0.25   0.      0.      0.25   0.      0.      0.      2.5
0.25    0.     0.25    0.      0.      0.25   0.      0.      2.5
0.      0.25   0.      0.25    0.      0.      0.25   0.      2.5
0.      0.     0.25    0.      0.      0.      0.      0.25   2.5
0.25    0.     0.      0.      0.      0.25   0.      0.      0.
0.      0.25   0.      0.      0.25   0.      0.25   0.      0.
0.      0.     0.25    0.      0.      0.25   0.      0.25   0.
0.      0.     0.      0.25    0.      0.      0.25   0.      0.
```

M85   M86

```
0.      0.25   0.25   0.      0.      0.      0.      0.      2.5
0.25    0.     0.      0.25    0.      0.      0.      0.      2.5
0.25    0.     0.      0.25    0.25   0.      0.      0.      0.
0.      0.25   0.25   0.      0.      0.25   0.      0.      0.
0.      0.     0.25    0.      0.      0.25   0.25   0.      0.
0.      0.     0.      0.25    0.25   0.      0.      0.25   0.
0.      0.     0.      0.      0.25   0.      0.      0.25   0.
0.      0.     0.      0.      0.      0.25   0.25   0.      0.
```

M87

```
0.      0.25   0.      0.25    0.      0.      0.      0.      0.
0.25    0.     0.25    0.      0.25    0.      0.      0.      0.
0.      0.25   0.      0.      0.      0.25    0.      0.      0.
0.25    0.     0.      0.      0.25    0.      0.25    0.      0.
0.      0.25   0.      0.25    0.      0.25    0.      0.25    0.
0.      0.     0.25    0.      0.25    0.      0.      0.      0.
0.      0.     0.      0.25    0.      0.      0.      0.25    0.25
0.      0.     0.      0.      0.25    0.      0.25    0.      0.
0.      0.     0.      0.      0.      0.      0.25    0.      0.
```

M88
0.2954	0.	0.0034	0.0693	0.0008	0.0083	0.0116
0.0002	0.0615	0.0109	0.0369	0.0006	0.0068	0.0023
0.0116	0.0005	0.0001	0.0021	0.0330	0.0295	0.0191
0.1158	0.0794	0.383	0.3876	0.0770	0.0867	0.0857
0.0186	0.0291	0.0322	0.0358	0.0543	0.0213	0.0555
0.1166	0.1552	0.1474	0.0802	0.1115	0.173	0.0371
0.0207	0.1066	0.0055	0.0387	0.0545	0.0369	0.0200

M89
```
   8722.
    919.
  56836.
 153806.
  18166.
  17750.
  31382.
```

M90 M91
1.	-1.	-1.	0.	0.	0.	0.	0.	0.	0.	0.
1.	0.	0.	1.	0.	-1.	0.	0.	0.	0.	0.
0.	0.	1.	0.	-1.	0.	0.	0.	0.	-1.	0.
0.	1.	0.	1.	1.	0.	-1.	-1.	0.	0.	0.
0.	0.	0.	0.	0.	1.	-1.	0.	-1.	0.	0.
2.	6.	0.	-4.	0.	0.	0.	0.	0.	0.	0.
0.	6.	-3.	0.	-3.	0.	0.	0.	0.	0.	0.
0.	0.	0.	0.	0.	0.	2.	-6.	-2.	0.	0.
0.	0.	0.	0.	3.	0.	0.	6.	0.	0.	0.
0.	0.	0.	4.	0.	0.	2.	0.	0.	0.	20.

M101
2.	0.	0.	0.	-1.	0.
4.	2.	0.	0.	-2.	-1.
0.	1.	0.	-1.	0.	0.
0.	0.	1.	0.	0.	-2.
-2.	0.	1.	-2.	0.	0.

M102
2.	0.	0.	0.	-1.	0.
4.	4.	0.	0.	-2.	-1.
0.	1.	0.	-1.	0.	0.
0.	0.	1.	0.	0.	-2.
-2.	-1.	1.	-2.	0.	0.

M103

2.	0.	0.	-1.	0.	0.
7.	0.	0.	0.	0.	-1.
0.	1.	0.	0.	-2.	0.
0.	0.	1.	0.	0.	-2.
-2.	-1.	1.	-3.	0.	0.

M104

1.	0.	0.	0.	-1.
3.	11.	-2.	-1.	0.
0.	12.	-1.	0.	0.
0.	22.	0.	-2.	0.

M105

1.	0.	0.	-1.	0.	0.	0.
1.	0.	0.	0.	-1.	0.	0.
0.	1.	0.	0.	0.	-1.	0.
0.	3.	0.	0.	0.	-1.	-1.
0.	0.	1.	0.	0.	0.	-2.
0.	-1.	1.	-2.	0.	0.	0.

M106

1.	0.	0.	-1.	0.	0.	0.
2.	6.	1.	-1.	0.	-3.	-2.
1.	0.	1.	-3.	0.	0.	-1.
0.	1.	0.	0.	-1.	0.	0.
0.	2.	0.	0.	0.	-1.	0.
0.	1.	-1.	1.	0.	0.	0.

M107

2.	0.	1.	0.	-2.	0.
2.	4.	0.	0.	-1.	-2.
0.	1.	0.	-1.	0.	0.
0.	-2.	1.	-2.	0.	0.

M108

2.	0.	0.	-1.	-1.	-1.	0.	0.
6.	2.	0.	0.	-1.	-1.	0.	-1.
1.	1.	0.	-2.	-2.	0.	0.	0.
0.	0.	3.	0.	0.	-3.	-1.	0.
0.	0.	-1.	0.	1.	0.	1.	-1.

Page 128

```
M109                                                       M110
   1.   -1.    0.    0.    0.    0.    0.    0.     -50.
   1.    0.   -1.    0.   -1.    0.    0.    0.     -30.
   0.    1.   -1.   -1.    0.    1.    0.    0.       0.
   0.    0.    0.    1.    0.    0.    1.    0.      80.
   0.    0.    0.    0.    1.    0.    0.    1.      20.
   0.    0.    0.    0.    0.    1.    1.   -1.      40.
M111
   0.    1.    1.    1.    0.    0.    0.    0.    0.
   1.    0.    0.    1.    1.    0.    0.    0.    0.
   1.    0.    0.    1.    0.    1.    0.    0.    0.
   1.    1.    1.    0.    1.    0.    1.    1.    0.
   0.    1.    0.    1.    0.    0.    0.    1.    1.
   0.    0.    1.    0.    0.    0.    1.    0.    0.
   0.    0.    0.    1.    0.    1.    0.    1.    0.
   0.    0.    0.    1.    1.    0.    1.    0.    1.
   0.    0.    0.    0.    1.    0.    0.    1.    0.
M112
   0.    1.    1.    0.    1.    0.    0.    0.    0.    0.
   0.    0.    1.    1.    1.    0.    0.    0.    0.    0.
   0.    0.    0.    1.    0.    1.    0.    0.    0.    0.
   0.    0.    0.    0.    0.    0.    0.    0.    0.    0.
   0.    0.    0.    0.    0.    1.    0.    1.    0.    0.
   0.    0.    0.    0.    0.    0.    0.    0.    1.    1.
   0.    0.    0.    0.    0.    0.    0.    1.    1.    0.
   0.    0.    0.    0.    0.    0.    0.    0.    1.    1.
   0.    0.    0.    0.    0.    0.    0.    0.    0.    1.
   0.    0.    0.    0.    0.    0.    0.    0.    0.    0.
M121
   1.         1.         1.
   0.         1.         3.
  -1.        -2.         2.
M122
   1.         1.         0.         0.
   0.        -3.         3.         0.
   2.         0.         0.         6.
   5.         3.         2.         0.
```

M123

3.	-1.	7.
0.	-2.	2.
1.	3.	-1.

M124

1.	3.	13.	-5.
2.	-4.	-4.	5.
-3.	5.	3.	6.

M125

4.	0.	1.	1.	0.
3.	0.	0.	1.	-1.
0.	1.	2.	0.	1.
5.	-2.	1.	0.	3.

M126

2.	1.	7.	17.
5.	0.	9.	0.
-1.	-6.	-3.	-6.
3.	-5.	4.	-7.

M127

1.	1.	1.	2.	4.
-3.	0.	3.	1.	-5.
1.	-2.	-5.	-5.	-3.
7.	4.	1.	11.	25.

M128

1.	1.	5.	0.	-2.
1.	0.	-3.	6.	5.
0.	1.	2.	0.	-1.
1.	0.	7.	-5.	-5.

M129

1.	-3.	-3.
4.	2.	0.
-1.	1.	3.

M130

3.	-2.	1.
5.	0.	-1.
4.	1.	2.

M131

1.	1.	1.	-1.
1.	1.	-1.	1.
1.	-1.	1.	1.
-1.	1.	1.	1.

M132

1.	3.	0.	-5.
2.	1.	3.	4.
-1.	-1.	2.	-7.
0.	2.	1.	2.

Page 130

M133

1.	0.	0.	0.
0.	1.	0.	0.
0.	0.	1.	0.
1.	1.	1.	1.

M134

1.	1.	0.	0.
0.	0.	1.	1.
1.	-1.	0.	0.
0.	0.	1.	-1.

M135

1.	0.	0.	0.	0.
4.	1.	0.	0.	0.
6.	3.	1.	0.	0.
4.	3.	2.	1.	0.
1.	1.	1.	1.	1.

M136

1.	2.	-3.	-4.
-8.	0.	-2.	1.
6.	-1.	1.	-1.
10.	5.	4.	0.

M137

1.	0.	1.
0.	0.	-1.
0.	1.	0.
-1.	2.	2.

M138

-3.	0.
4.	1.
0.	2.
0.	-2.

M139

-3.	0.	1.
1.	4.	0.
1.	1.	0.
2.	5.	0.

M140

-3.	-1.	0.	1.
1.	3.	4.	0.
1.	1.	1.	0.
2.	4.	5.	0.

M141

-3.	1.	-1.	0.	1.
1.	1.	3.	4.	0.
1.	0.	1.	1.	0.
2.	1.	4.	5.	0.

M142

2.	0.	0.	0.
1.	2.	0.	0.
0.	0.	0.	1.
0.	0.	-6.	5.

M143

0.	1.	1.
3.	0.	-1.
-2.	2.	0.
4.	1.	0.

M144

-1.
4.
-5.
1.

M145

2.	6.	1.
0.	-2.	-1.
-1.	2.	-2.
2.	4.	0.

M146

4.
-4.
-4.
1.

M147

1.	-4.	0.
-3.	0.	3.
0.	1.	-3.
2.	-2.	1.
3.	1.	0.

M148

0.
7.
-1.
0.
2.

M149

-4.	1.	-2.
0.	-3.	-6.
1.	0.	1.
-2.	2.	2.
1.	3.	7.

M150

3.	-8.
3.	4.
-1.	-5.
0.	1.
-4.	-7.

M151

1.	1.	1.	1.
0.	1.	3.	0.
-1.	-2.	2.	4.

M152

64.	141.
66.	148.
68.	155.
70.	163.
72.	172.
74.	182.
76.	190.
78.	198.

M153

1.	0.0036630	1.52170
1	0.0035336	2.22029
1.	0.0034130	2.86448
1.	0.0033003	3.46010
1.	0.0031949	4.01277
1.	0.0030960	4.52721

M154

2.	86.6
3.	85.6
4.	85.3
5.	77.3
6.	73.7
7.	72.
8.	70.3
9.	66.2
10.	66.1

M155

14.5	76.	31.4
14.2	75.	29.9
15.7	85.	40.8
14.4	76.	31.
16.8	71.	35.5
20.5	82.	65.8
14.1	80.	32.7
14.8	69.	31.3
18.9	72.	48.4
14.	66.	25.5

M156

1431.3	1298.9
1493.2	1337.7
1551.3	1405.9
1599.8	1456.7
1668.1	1492.
1728.4	1538.8
1797.4	1621.9
1916.3	1689.6
1896.6	1674.
1931.7	1711.9
2001.	1803.9
2066.6	1883.8
2167.4	1961.
2212.6	2004.4
2214.3	2000.4
2248.6	2024.2
2261.5	2050.7
2331.9	2146.
2470.6	2246.3
2528.	2324.5

M157

10.	96.1	4.5654
60.	80.	4.382
95.	69.8	4.2456
145.	57.9	4.0587
295.	32.2	3.472
590.	11.	2.3979

M158

14.	0.	17306.4
15.	0.	17408.0
16.	0.	17506.7
17.	0.	17603.8
13.	1.	16991.1
14.	1.	17092.5
15.	1.	17193.9
16.	1.	17290.9
10.	2.	16463.4
11.	2.	16566.0
12.	2.	16672.9
13.	2.	16779.5

M161

1.	0.	1.	-1.
0.	-1.	2.	2.
0.	2.	2.	2.
0.	-3.	-6.	-6.

M162

4.	-1.	1.
-1.	4.	-1.
1.	-1.	4.

M163

-1.	-1.	1.	-1.
0.	-3.	4.	0.
0.	0.	1.	0.
-4.	2.	-2.	-1.

M164

-1.	-1.	2.	-3.
4.	3.	0.	6.
0.	0.	5.	9.
0.	0.	-1.	-1.

M165

0.	-6.	6.
11.	-18.	7.
5.	-6.	1.

M166

0.85	0.15	0.05
0.05	0.75	0.05
0.1	0.1	0.9

M167

0.5	0.33	0.	0.	0.	0.
0.5	0.33	0.33	0.	0.	0.
0.	0.33	0.33	0.33	0.	0.
0.	0.	0.33	0.33	0.33	0.
0.	0.	0.	0.33	0.33	0.5
0.	0.	0.	0.	0.33	0.5

M168

0.	0.33	0.25	0.33	0.	0.	0.
0.33	0.	0.25	0.	0.5	0.	0.
0.33	0.33	0.	0.33	0.	0.33	0.
0.33	0.	0.25	0.	0.	0.	0.5
0.	0.33	0.	0.	0.	0.33	0.
0.	0.	0.25	0.	0.5	0.	0.5
0.	0.	0.	0.33	0.	0.33	0.

M169

1.	0.4	0.	0.	0.	0.
0.	0.25	0.4	0.	0.	0.
0.	0.35	0.25	0.4	0.	0.
0.	0.	0.35	0.25	0.4	0.
0.	0.	0.	0.35	0.25	0.
0.	0.	0.	0.	0.35	1.

M170

0.	2.	2.	0.
0.38	0.	0.	0.
0.	0.3	0.	0.
0.	0.	0.25	0.

M171

0.	0.259	1.099	0.699	0.154	0.	0.
0.953	0.	0.	0.	0.	0.	0.
0.	0.994	0.	0.	0.	0.	0.
0.	0.	0.99	0.	0.	0.	0.
0.	0.	0.	0.983	0.	0.	0.
0.	0.	0.	0.	0.963	0.	0.
0.	0.	0.	0.	0.	0.915	0.

M172

0.	0.022	0.532	0.209	0.009	0.	0.
0.99	0.	0.	0.	0.	0.	0.
0.	0.997	0.	0.	0.	0.	0.
0.	0.	0.996	0.	0.	0.	0.
0.	0.	0.	0.993	0.	0.	0.
0.	0.	0.	0.	0.98	0.	0.
0.	0.	0.	0.	0.	0.952	0.

M173

0.	0.196	0.687	0.107	0.006	0.	0.
0.978	0.	0.	0.	0.	0.	0.
0.	0.996	0.	0.	0.	0.	0.
0.	0.	0.994	0.	0.	0.	0.
0.	0.	0.	0.99	0.	0.	0.
0.	0.	0.	0.	0.976	0.	0.
0.	0.	0.	0.	0.	0.938	0.

M174

1.	0.25	0.	0.	0.	0.
0.	0.25	0.	1.	0.25	0.
0.	0.	0.	0.	0.25	0.
0.	0.25	0.	0.	0.	0.
0.	0.25	1.	0.	0.25	0.
0.	0.	0.	0.	0.25	1.

M175

-0.08	0.	0.03
0.04	-0.04	0.
0.04	0.04	-0.08

M176

-0.075	0.	0.025	0.
0.025	-0.0333	0.00833	0.
0.	0.	-0.0333	0.0333
0.05	0.0333	0.	-0.0833

M177

0.	1.	0.	0.
-10.	0.	4.	0.
0.	0.	0.	1.
4.	0.	-4.	0.

M178

-2.	1.	0.	0.
1.	-2.	0.	0.
1.	1.	-5.	0.
0.	0.	5.	0.

M179

-0.03	0.04	0.	0.
0.03	-0.08	0.03	0.
0.	0.04	-0.06	0.04
0.	0.	0.03	-0.04

Appendix IV

The MAX Diskette

STARTING MAX

The MAX diskette as you originally receive it cannot be used to start your computer, but you may start MAX by the following steps:

1. Start your computer with DOS, according to the instructions for your computer.
2. Insert the MAX diskette into the A: disk drive.
3. Type **MAX** and press the RETURN (or Enter) key.
   If that does not work, try typing **A:MAX**.

You should then see MAX's name and a copyright notice, followed by the **Command:** prompt (see the Overview). However, you will probably find that MAX's **edit** command does not work properly. See below for instructions on how to correct this.

If you wish to be able to start your computer and MAX all in one step, you may transfer DOS to your MAX diskette. This need only be done once. The first two steps are the same as above: Start your computer and then insert the MAX diskette into the A: drive. The next step is to type **A:setup**. (However, see the next paragraph if you have a hard disk.) This command will ask you to insert your DOS diskette in the B: drive and then will transfer DOS to your MAX diskette. (It should also enable the **edit** command to work properly; see below.) From then on, you should be able to get directly into MAX by starting your computer with this modified MAX diskette; the steps described in the first paragraph above will no longer be needed.

If you have a hard disk, you may wish to transfer DOS from your hard disk or some other drive besides the B: drive. (This is especially recommended if you have only one disk drive.) In that case, enter the name of that drive, followed by a colon, after the **setup** command. For example, to install DOS from your hard disk C: you would type **A:setup C:** in place of **A:setup** as stated in the previous paragraph. For this to work, however, you must first have set the current DOS default directory on the C: drive (or whatever drive you are installing DOS from) to the directory which contains your DOS system files. (If they are in the root directory, you need not do anything extra; if you do not know how to use directories, try using the B: drive as described above.)

THE EDIT COMMAND

Unlike the rest of MAX, the **edit** command relies on special capabilities of your computer to control what you see on the screen. If you use the **setup** command as described above and thereafter start your computer with the MAX diskette, the **edit** command should work properly. If it still does not, the following information may help. **Edit** uses ANSI escape sequences for screen control. In most versions of MS-DOS, these require the use of a small program called ANSI.SYS which should be on your original DOS diskette if your MS-DOS requires it. In order to use ANSI.SYS, the disk from which you start your computer must contain (in its root directory) a file called CONFIG.SYS containing the line **device = ansi.sys**. The file ACONFIG.SYS provided on the MAX diskette contains this line, and if your DOS diskette contains ANSI.SYS then the **setup** command will transfer ANSI.SYS to your MAX diskette and rename ACONFIG.SYS to CONFIG.SYS. If you use a different disk to start your computer, or if **setup** could not locate ANSI.SYS in order to transfer it (in which case you will be informed by a message on the screen at the time), you may need to see to this yourself. For further information about CONFIG.SYS and ANSI.SYS, refer to your DOS manual, especially the section about "configuring your system."

Note that even if **edit** does not work for you, this is not a serious deficiency; all the other MAX commands will still work properly, and the **change** command can substitute for the **edit** command.

THE FILES ON THE MAX DISKETTE

`MAX.EXE`   This is the MAX program itself.

`GRP00.DAT`   The nine `.DAT` files contain the group storage matrices.
`GRP01.DAT`
`GRP02.DAT`
`GRP03.DAT`
`GRP04.DAT`
`GRP05.DAT`
`GRP06.DAT`
`GRP07.DAT`
`GRP08.DAT`

`HLPALG.HLP`   The eight `.HLP` files are the help files containing the
`HLPCON.HLP`   information you see when you use MAX's **help** command.
`HLPCRE.HLP`
`HLPFIL.HLP`
`HLPINF.HLP`
`HLPMOD.HLP`
`HLPNUM.HLP`
`HLPOVE.HLP`

`DEMO1.MAC`
`DEMO2.MAC`
`DEMO3.MAC`
`DEMO4.MAC`
`LS.MAC`
`LSPARAM.MAC`
	The `.MAC` files are sample script files for use with MAX's **demonstrate** or **execute** commands

`SETUP.BAT`
`AUTOEXEC.BAT`
`ACONFIG.SYS`

`SETUP.BAT` is a batch file which performs the **setup** command as described in the first part of this appendix. Once it has been used to transfer DOS to the MAX diskette, then `AUTOEXEC.BAT` starts up MAX automatically when the computer is started from this diskette. `ACONFIG.SYS` is used by **setup**; see the previous section of this appendix for more information.

`MATRIX.BAS`

This is a BASIC program for creating a matrix whose entries are defined by an arbitrary function or subroutine. The matrix is written into a file from which it can be brought into MAX by the **retrieve** command. To define your own matrix, you must edit the program to change the subroutine that defines the entries. (For more information on this, read the comments in the program.) When you run the program, it will ask for a file name and the dimensions of the matrix to be written to that file. The file name has a default extension of `.MAT`, just as in the **save** command. For information about BASIC programs, refer to your BASIC manual.

REFERENCES

Arganbright, Deane E. *Mathematical Applications of Electronic Spreadsheets.* McGraw-Hill, 1985.

Dongarra, J. J., C. B. Moler, J. R. Bunch, and G. W. Stewart. *LINPACK Users' Guide.* SIAM, 1979.

Helzer, Garry. *Applied Linear Algebra with APL.* Little, Brown & Company, 1983.

Herman, Eugene A., Howard Anton, Alan Tucker, and Garry Helzer. "The use of computing in the teaching of linear algebra" in *Computing and Mathematics: The Use of Computers in Undergraduate Instruction* by David A. Smith et al (MAA Notes Series). Math. Assoc. of Amer. (to appear in 1987).

Rorres, Chris and Howard Anton. *Applications of Linear Algebra, 3rd ed.* John Wiley & Sons, 1984.

Smith, B. T. et al. *Matrix Eigensystems Routines - EISPACK Guide, 2nd ed.* Springer-Verlag, 1976.

Stewart, Gilbert W. *Introduction to Matrix Computations.* Academic Press, 1973.

Strang, Gilbert. *Linear Algebra and its Applications, 2nd ed.* Academic Press, 1980.

Tucker, Alan. *A Unified Introduction to Linear Algebra.* Macmillan, 1987.

Wilkinson, James H. and C. Reinsch. *Handbook for Automatic Computation, vol II, Linear Algebra.* Springer-Verlag, 1971.

Index of Symbols

Index